MW00678301

DATA-VARIANT KERNEL ANALYSIS

Wiley Series on
Adaptive and Cognitive Dynamic Systems

Editor: Simon Haykin

A complete list of titles in this series appears at the end of this volume.

DATA-VARIANT KERNEL ANALYSIS

Yuichi Motai, Ph.D.
Sensory Intelligence Laboratory,
Department of Electrical and Computer Engineering,
Virginia Commonwealth University
Richmond, VA

Published by John Wiley & Sons, Inc., Hoboken, New Jersey
Published simultaneously in Canada

Library of Congress Cataloging-in-Publication Data:

Motai, Yuichi.
Data-variant kernel analysis / Yuichi Motai.
pages cm. – (Wiley series on Adaptive and Cognitive Dynamic Systems)
Includes bibliographical references and index.
ISBN 978-1-119-01932-9 (hardback)
1. Kernel functions. 2. Big data–Mathematics. I. Title.
QA353.K47M68 2015
515′.9–dc23

2015000041

Cover image courtesy of iStockphoto © Hong Li

Typeset in 10/12pt TimesLTStd by Laserwords Private Limited, Chennai, India

Printed in the United States of America

10 9 8 7 6 5 4 3 2 1

1 2015

The author dedicates this book in memoriam to his father, Osami Motai, who passed away on June 29, 2013.

CONTENTS

Chapter 2 Offline Kernel Analysis 41

Chapter 3 Group Kernel Feature Analysis 69

Chapter 4 Online Kernel Analysis 97

Chapter 5 Cloud Kernel Analysis 121

LIST OF FIGURES

LIST OF TABLES

PREFACE

Kernel methods have been extensively studied in pattern classification and its applications for the past 20 years. Kernel may refer to diverse meanings in different areas such as Physical Science, Mathematics, Computer Science, and even Music/Business. For the area of Computer Science, the term "kernel" is used in different contexts (i) central component of most operating systems, (ii) scheme-like programming languages, and (iii) a function that executes on OpenCL devices. In machine learning and statistics, the term kernel is used for a pattern recognition algorithm. The kernel functions for pattern analysis, called kernel analysis (KA), is the central theme of this book. KA uses "kernel trick" to replace feature representation of data with similarities to other data. We will cover KA topics ranging from the fundamental theory of kernel functions to applications. The overall structure starts from Survey in Chapter 1. On the basis of the KA configurations, the remaining chapters consist of Offline KA in Chapter 2, Group KA in Chapter 3, Online KA in Chapter 4, Cloud KA in Chapter 5, and Predictive KA in Chapter 6. Finally, Chapter 7 concludes by summarizing these distinct algorithms.

Chapter 1 surveys the current status, popular trends, and developments on KA studies, so that we can oversee functionalities and potentials in an organized manner:

- Utilize KA with different types of data configurations, such as offline, online, and distributed, for pattern analysis framework.
- Adapt KA into the traditionally developed machine learning techniques, such as neural networks (NN), support vector machines (SVM), and principal component analysis (PCA).
- Evaluate KA performance among those algorithms.

Chapter 2 covers offline learning algorithms, in which KA does not change its approximation of the target function, once the initial training phase has been absolved. KA mainly deals with two major issues: (i) how to choose the appropriate kernels for offline learning during the learning phase, and (ii) how to adopt KA into the traditionally developed machine learning techniques such as NN, SVM, and PCA, where the (nonlinear) learning data-space is placed under the linear space via kernel tricks.

Chapter 3 covers group KA as a data-distributed extension of offline learning algorithms. The data used for Chapter 3 is now extended into several databases. Group KA for distributed data is explored to demonstrate the big-data analysis with the comparable performance of speed and memory usages.

Chapter 4 covers online learning algorithms, in which KA allows the feature space to be updated as the training proceeds with more data being fed into the algorithm. The feature space update can be incremental or nonincremental. In an incremental update, the feature space is augmented with new features extracted from the new data, with a possible expansion to the feature space if necessary. In a nonincremental update, the dimension of the feature space remains constant as the newly computed features may replace some of the existing ones. In this specific chapter, we also identify the following possibilities of online learning:

- Synthesize offline learning and online learning using KA, which suggests other connections and potential impact both on machine learning and on signal processing.
- Extend KA with different types of data configurations, from offline to online for pattern analysis framework.
- Apply KA into practical learning setting, such as biomedical image data.

Chapter 5 covers cloud data configuration. The objective of this cloud network setting is to deliver an extension of distributed data. KA from offline and online learning aspects are carried out in the cloud to give more precise treatments to nonlinear pattern recognition without unnecessary computational complexity. This latest trend of big-data analysis may stimulate the emergence of cloud studies in KA to validate the efficiency using practical data.

Chapter 6 covers longitudinal data to predict future state using KA. A time-transitional relationship between online learning and prediction techniques is explored, so that KA can be applied to adaptive prediction from online learning. The prediction performance over different time periods is evaluated in comparison to KA alternatives.

Chapter 7 summarizes these distinct data formations used for KA. The data handling issues and potential advantages of data-variant KAs are listed. The supplemental material includes MATLAB® codes in Appendix.

The book is not chronological, and therefore, the reader can start from any chapter. All the chapters were formed by themselves and are relevant to each other. The author has organized each chapter assuming the readers had not read the other chapters.

ACKNOWLEDGMENTS

This study was supported in part by the School of Engineering at Virginia Commonwealth University and the National Science Foundation.

The author would like to thank his colleagues for the effort and time they spent for this study:

Dr. Hiroyuki Yoshida for providing valuable colon cancer datasets for experimental results.

Dr. Alen Docef for his discussion and comments dealing with joint proposal attempts.

The work reported herein would not have been possible without the help of many of the past and present members of his research group, in particular:

Dr. Awad Mariette, Lahiruka Winter, Dr. Xianhua Jiang, Sindhu Myla, Dr. Dingkun Ma, Eric Henderson, Nahian Alam Siddique, Ryan Meekins, and Jeff Miller.

1

SURVEY[1]

1.1 INTRODUCTION OF KERNEL ANALYSIS

Kernel methods have been widely studied for pattern classification and multidomain association tasks [1–3]. Kernel analysis (KA) enables kernel functions to operate in the feature space without ever computing the coordinates of the data in that space, but rather by simply computing the inner products between the images of all pairs of data in the feature space [4, 5]. This operation is often less computational than the explicit computation of the coordinates [3, 6, 7]. This approach is called the kernel trick. Kernel functions have been introduced for sequence data, graphs, text, images, as well as vectors [8–14].

Kernel feature analysis attracts significant attention in the fields of both machine learning and signal processing [10, 15]; thus there are demands to cover this state-of-the-art topic [16]. In this survey, we identify the following popular trends and developments in KA, so that we can visualize the merits and potentials in an organized manner:

- Yield nonlinear filters in the input space to open up many possibilities for optimum nonlinear system design.
- Adapt KA into the traditionally developed machine learning techniques for nonlinear optimal filter implementations.
- Explore kernel selection for distributed databases including solutions of heterogeneous issues.

[1]This chapter is a revised version of the author's paper in IEEE Transactions on Neural Networks and Learning Systems. DOI: 10.1109/TNNLS.2014.2333664, approved by IEEE Intellectual Property Rights.

Data-Variant Kernel Analysis, First Edition. Yuichi Motai.

Constructing composite kernels is an anticipated solution for heterogeneous data problems. A composite kernel is more relevant for the dataset and adapts itself by adjusting its composed coefficient parameters, thus allowing more flexibility in the kernel choice [3, 8, 17–20].

The key idea behind the KA method is to allow the feature space to be updated as the training proceeds with more data being fed into the algorithm [15, 21–26]. This feature space update can be incremental or nonincremental. In an incremental update, the feature space is augmented with new features extracted from the new data, with a possible expansion of the feature space if necessary [21–27]. In a nonincremental update, the dimension of the feature space remains constant, as the newly computed features may replace some of the existing ones [8, 19, 20]. In this survey, we also identify the following possibilities:

- A link between offline learning and online learning using KA framework, which suggests other connections and a potential impact on both machine learning and signal processing.
- A relationship between online learning and prediction techniques to merge them together for an adaptive prediction from online learning.
- An online novelty detection with KA as an extended application of prediction algorithms from online learning. These algorithms listed in this survey are capable of operating with kernels, including support vector machines (SVMs) [12, 28–34] Gaussian processes [35–38], Fisher's linear discriminant analysis (LDA) [19, 39], principal component analysis (PCA) [3, 9–11, 20, 22, 24–27, 40], spectral clustering [41–47], linear adaptive filters [48–53], and many others.

The objective of this survey is to deliver a comprehensive review of current KA methods from offline and online learning aspects [1, 19, 21]. Research studies on KA are carried out in the areas of computer science or statistics to give precise treatments to nonlinear pattern recognition without unnecessary computational complexity [4, 5, 28]. This latest survey of KA may stimulate the emergence of neural network studies in computer science applications and encourage collaborative research activities in relevant engineering areas [8–12].

We begin to describe the offline kernel methods by basic understanding of KA in Section 1.2. Then, this survey shows advanced distributed database technology for KA in Section 1.3. In the following sections, we point out online learning frameworks (vs offline learning in Section 1.2) using kernel methods in Section 1.4 and then the prediction of data anomaly in Section 1.5. Finally, we conclude with the future directions of KA in Section 1.6.

1.2 KERNEL OFFLINE LEARNING

Offline learning is a machine learning algorithm in which the system does not change its approximation of the target function, once the initial training phase has been absolved.

KA mainly deals with two major issues: (i) how to choose the appropriate kernels for offline learning during the learning phase, and (ii) how to adopt KA into the traditionally developed machine learning techniques such as neural network (NN), SVM, and PCA, where the nonlinear learning data space is placed under the linear space via kernel tricks.

1.2.1 Choose the Appropriate Kernels

Selecting the appropriate kernels is regarded as the key problem in KA, as it has a direct effect on the performance of machine learning techniques [35, 54–65]. The proper kernels are identified by *priori* without the analysis of databases, and the kernel selection is often implemented in the initial stage of offline learning. Commonly used kernel functions are as follows [20, 56, 66, 67]:

The linear kernel: $K(x, x_i) = x^T x_i$
The polynomial kernel: $K(x, x_i) = (x^T x_i + \text{Offset})^d$
The Gaussian radial basis function (RBF) kernel: $K(x, x_i) = \exp(-\|x - x_i\|^2/2\sigma^2)$
The exponential Gaussian RBF kernel: $K(x, x_i) = \exp(-\|x - x_i\|/2\sigma^2)$
The Laplace kernel: $K(x, x_i) = \exp(-\sqrt{\|x - x_i\|^2/\sigma^2})$
The Laplace RBF kernel: $K(x, x_i) = \exp(-\sigma\|x - x_i\|)$
The sigmoid kernel: $K(x, x_i) = \tanh(\beta_0 x^T x_i + \beta_1)$
The analysis of variance (ANOVA) RB kernel: $K(x, x_i) = \sum_{k=1}^{n} \exp(-\sigma(x^k - x_i^k)^2)^d$
The Cauchy kernel: $K(x, x_i) = 1/(1 + \|x - x_i\|^2/\sigma)$
The multiquadric kernel: $K(x, x_i) = \sqrt{\|x - x_i\|^2 + \sigma^2}$
The power exponential kernel: $K(x, x_i) = \exp((-\|x - x_i\|^2/2\sigma^2)^d)$
The linear spline kernel: $K(x, x_i) = 1 + xx_i \min(x, x_i) - ((x + x_i)/2)(\min(x, x_i)^2 + (\min(x, x_i)^3/3))$.

Many KA studies do not address appropriate kernel functions; instead, they simply choose well-known traditional kernels empirically by following other studies [56, 57, 68–71].

As shown in Table 1.1, there are four proposed kernel methods for choosing proper kernels: (i) linear combinations of base kernel functions, (ii) hyperkernels, (iii) difference of convex (DC) functions programming, and (iv) convex optimization. The representative accuracy is calculated as the average of the maximum accuracy with its standard deviation for each kernel selection approach [(i)–(iv)], according to the papers listed. On the basis of Table 1.1, the DC approach consistently achieved the best accuracy, using a 95% level of confidence.

Linear Combinations of Base Kernel Functions The linear combination of base kernel functions approach is based on the fact that the combination of kernels satisfying the Mercer's condition can generate another kernel [73]. In other

Table 1.1 Kernel selection approaches and related research

Kernel selection approach	Relevant studies	Representative accuracy
Linear combinations	[58, 60, 72, 73]	93.6 ± 8.4
Hyperkernels	[35, 57, 61–63, 74]	95.9 ± 4.3
Difference of convex (DC) functions	[36, 64, 75, 76]	99.1 ± 0.76
Convex optimization	[69, 71, 77, 81]	96.5 ± 4.1

words, the appropriate kernel is chosen by estimating the optimal weighted combination of base kernels [56, 67], to satisfy the Mercer's condition $\iint K(x, x_i)g(x)g(x_i)\, dxdx_i \geq 0$.

The proper kernels [58, 59, 68, 72, 73] are decided when base kernels $k_1, \ldots, k_D$ are given from Equation 1.1 [60, 73]:

$$K(x, \bar{x}) = \sum_{d=1}^{D} w_d k_d(x, \bar{x}) K \tag{1.1}$$

where D is the number of base kernels, and each kernel function $k_d : \chi \times \chi \rightarrow \mathbf{R}$ has two inputs and produces a scalar as the output. In addition, $w_1, \ldots, w_D$ are non-negative weights [60, 73]. As shown in Equation 1.1, this method is effective because weights w_d of useless kernels k_d are ideally set to zero.

Hyperkernels Hyperkernels are proposed by Ong et al. [61]; it can be used for kernel learning for the inductive setting [57]. The definition of hyperkernels is as follows [35, 57, 61–63, 74]:

Let x be a nonempty set, $\hat{x} = x \times x$ and $K : x \times x \rightarrow R$ with $K_x(\cdot) = K(\hat{\chi}, \cdot) = K(\cdot, \hat{\chi})$. Then, K is called a hyperkernel on x if and only if

1. K is a positive definite on $\hat{x}$ and
2. For any $\hat{\chi} \in \hat{x}$, K_x is a positive definite on x.

These hyperkernels are flexible in kernel learning for a vast range of data sets, as they have the invariant property of translation and rotation simultaneously [63]. In other studies [35, 57, 62, 63, 74], these hyperkernels are extended into a variety of aspects as in Table 1.2.

These hyperkernel approaches listed in Table 1.2 have somehow overlapped the principles and advantages. These extensions include a method using customized optimization for measuring the fitness between a kernel and the learning task.

DC Functions-Programming Algorithm The kernel is selected by specifying its objective function [58, 61, 76–80] by optimizing a predefined family of kernels. DC-programming is one of the algorithms for those optimization problems, expressed

Table 1.2 Hyperkernel method comparison

Method	Principle	Advantage
Iterative optimization [57]	Second-order cone programming	Efficient calculation
Invariance [63]	Invariant translation and rotation simultaneously	Flexible wide range of data sets
Learning [62]	Learning hyperplane	Optimal parameters
Gaussian [35]	Wishart hyperkernels	Less computational complexity
Structural risk minimization [74]	Regularization penalty into the optimization	Low bias, high variance, prevent overfitting

as Equation 1.2 [64, 75]:

$$\min D(\rho) = \min(g(\rho) - h(\rho)),$$
$$g(\rho) = \frac{1}{n_1^2 \mathbf{1}^T \mathbf{K}_{1,1} \mathbf{1}} + \frac{1}{n_2^2 \mathbf{1}^T \mathbf{K}_{2,2} \mathbf{1}} \text{ and } h(\rho) = \frac{2}{n_1 n_2 \mathbf{1}^T \mathbf{K}_{1,2} \mathbf{1}} \quad (1.2)$$

where $g(\rho)$ and $h(\rho)$ are convex functions, $\mathbf{1} = \{1, ..1\}^T$, n is the number of data, and $[\mathbf{K}]_{i,j} = k(x_i, x_j)$ for either class 1 or 2. This means the difference of convex functions $D(\rho)$ can be a nonconvex function. Thus, the DC-programming algorithm can be applied to both nonconvex and convex problems [64, 68, 75, 36].

To solve the nonconvex problem, Neumann et al. [75] adopt DC functions, which minimize the difference between convex functions as the kernel choice method. Argyriou et al. [68] propose the DC-programming algorithm on the basis of mini-max optimization problems involved in the greedy algorithm. This study [68] has a significant advantage, as choosing the kernel among basic kernels in advance is not required.

Convex Optimization Algorithm The problem for choosing the kernel can be reformulated as a manageable, convex optimization problem [69, 71, 77, 81]. Kim et al. [69] and Khemchandani et al. [71] address the kernel selection in the kernel Fisher discriminant analysis (KFDA) to maximize a Fisher discriminant ratio (FDR). The basic concept of adopting convex optimization is Equation 1.3 as follows [69, 71]:

$$\max F_\lambda^*(K) \quad (1.3)$$

subject to $K = \theta K_1 + (1 - \theta) K_2$, where kernel functions ${}^\exists K_1, {}^\exists K_2 \in \kappa$, *and* $\forall \theta \in [0,\ 1]$. F_λ^* is FDR with a positive regularization parameter λ, and κ is a set of kernel functions K.

Other kernel-based methods select the kernel optimizing over the Gram matrix with a fixed weight vector, which has similar computational complexity but cannot ensure that the globally optimal kernel is always found. In the case of the convex, the global optimization is achieved [69].

1.2.2 Adopt KA into the traditionally developed machine learning techniques

KA methods are used for the nonlinear extension to make it more applicable in addition to the existing offline methods, such as (i) neural network, (ii) SVM, and (iii) PCA.

Neural Network NN is a very successful technique, which uses input/output components that are trained to perform nonlinear classification [82]. This structured approach allows KA to be incorporated within NN in different ways [37, 83–96].

Some studies [37, 83–87] embed KA inside of the normal NN structure. A common example of combining the NN with KA is the radial basis function neural network (RBF-NN) [97]. This technique successfully uses a RBF kernel algorithm for the activation functions of the hidden layer in the form (1.4).

$$\eta(x, w) = \sum_{i=1}^{S1} \kappa_i(\|x - x_i\|) \tag{1.4}$$

where η is output, x is input with the hidden layer κ (Fig. 1.1).

Other nontraditional KA used in the general structure of the NN can be found in [88–92]. Although commonly standalone, NN techniques such as back-propagation and recurrence functions combine to train and expand the use of NN with KA in the literature [85, 86, 93–96]. Table 1.3 gives key references of back-propagation, recurrence, RBF, and some nontraditional techniques in conjunction with KA. On the basis of Table 1.3, many nontraditional approaches are proposed for reaching the best classification accuracy of each study listed in the relevant studies. The representative accuracy consists of the mean with 1 standard deviation among these relevant studies.

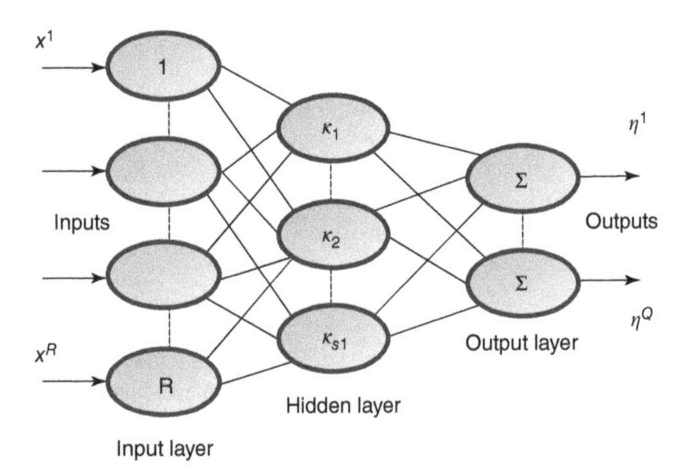

Figure 1.1 General RBF-NN [97]. The input and output layers perform functions similarly to a normal neural network. The hidden layer uses the RBF kernel, denoted as its activation function as shown in Equation 1.4.

Table 1.3 Neural network approaches related to kernel algorithm

Algorithms	Relevant studies	Representative accuracy
Back-propagation	[85, 86, 95, 96]	83.6 ± 10.8
Recurrence	[93, 94]	91.6 ± 10.5
Radial basis function (RBF)	[37, 83, 84, 86, 87]	90.0 ± 2.5
Nontraditional	[88–92]	96.7 ± 4.6

A comparison between KA and NN studies are conducted. Some NN [86, 94, 95] outperform non-KA-based NN. In the comparison of accuracy, more common KA techniques outperform standalone NN [98–101]. None of these references provides definitive evidence that KA or NN is generally better for all situations.

Support Vector Machine SVM originates from statistical learning theory developed by Vapnik [102], as a binary classifier. SVM is developed as a margin classifier [103]. The task in SVM is to find the largest margin separating the classifier hyperplane [104]. Equation 1.5 is an optimization problem that involves the cross-product of training data.

$$L(\alpha) = \sum_{i=1}^{m} \alpha_i - \frac{1}{2}\sum_{i,j=1}^{m} y_i y_j \alpha_i \alpha_j \mathbf{x}_i \cdot \mathbf{x}_j \tag{1.5}$$

where $\mathbf{x}$ is the input, α is a Lagrange multiplier, and y is the label for the distance criterion L. Equation 1.5 is a formulation of SVM for linearly separable data. The cross-product is also called linear kernel [105]. In order for SVM to work in non-separable (i.e., nonlinear) data, the data is mapped into a higher dimensional feature space, defined by the kernel function [52]. Kernel function is the inner product in the feature space and in the form of Equation 1.6:

$$K(x_i, x_j) = \Phi(x_i) \cdot \Phi(x_j) \tag{1.6}$$

Tolerance to error needs to be introduced. The SVM learning task then becomes to minimize (Eq. 1.7),

$$L(\alpha) = \sum_{i=1}^{m} \alpha_i - \frac{1}{2}\sum_{i,j=1}^{m} y_i y_j \alpha_i \alpha_j K(x_i, x_j) \tag{1.7}$$

subject to $0 \leq \alpha_i \leq C$ and $\sum \alpha y$ is a measure of tolerance to error. C is an additional constraint on the Lagrange multiplier. Equation 1.7 is a quadratic programming problem and can be solved using standard Quadratic Programming solver, called the C-SVM formulation [106]. A review on different methods to solve the optimization problem is discussed in [107]. Sequential minimal optimization (SMO) and other decomposition methods are used to solve the optimization problems listed in Table 1.4. This table shows SVM approaches consistently reach high classification accuracy.

Table 1.4 SVM approaches relevant to solve optimization problem

Algorithm	Relevant studies	Representative accuracy
Sequential minimal optimization (SMO)	[108, 109]	N.A.
Decomposition methods	[110, 111]	99.9 ± 0.0
Other methods	[112, 113]	92.3 ± 10.1

Among kernel functions used for SVM [37, 114], the three most common kernel functions used are polynomial, Gaussian, and sigmoid functions. SVM shows competitive performance for classification but does not outperform other techniques in regression problems [115]. Recent SVM developments introduce a new variant called twin support vector machine (TSVM) [116, 117], which uses nonparallel separating hyperplane unlike original formulation of SVM. Nonparallel hyperplanes are beneficial for preferential classification task, where one class is more significant than the others.

Principal Component Analysis PCA, also known as Karhunen–Loève transform (KLT) or proper orthogonal decomposition (POD), is a mathematical method for reducing dimension of feature space by representing in the orthogonal eigenvector space as shown in Fig. 1.2.

The transformation incorporates an eigenvector and eigenvalue calculation and is a linear transformation. The transformed space, in descending order of accounted variance, is orthogonal if the original dataset is jointly normally distributed. PCA is sensitive to relative scaling and expected value of the original feature. Being a linear model, PCA efficiently uses the nonlinearity provided by kernel association [118] in the form of higher dimension—making kernel principal component analysis (KPCA) a nonlinear component analysis tool. Bouveyron et al. [41] conduct an evaluation of PCA and KPCA in comparison to other similar methods for high-dimensional data clustering. In [119] PCA is generalized to use kernel for higher dimension input space. The developed supervised PCA as well as the supervised KPCA aims to find the

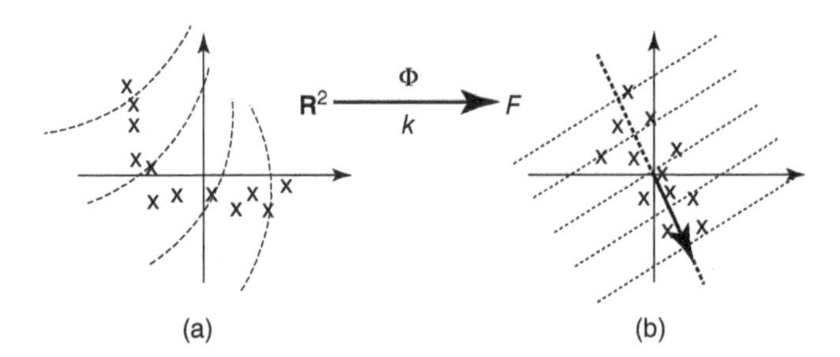

Figure 1.2 Principle of kernel PCA. (a) nonlinear in input space. (b) high-dimensional feature space corresponding to nonlinear input space. The contour lines of constant projections generate the principal eigenvectors. Kernel PCA does not actually require Φ to be mapped into F. All necessary computations are executed through the use of a kernel function k in input space $\mathbf{R}^2$.

principal components with maximum dependence on the response variables, which can be achieved by Equation 1.8.

$$\begin{aligned} &\underset{U}{\operatorname{argmax}} \quad \operatorname{tr}(U^T XHLHX^T U) \\ &\quad \text{subject to} \quad U^T U = \mathbf{I} \end{aligned} \tag{1.8}$$

where U is the optimal solution space, X is the input set, H is the Hilbert–Schimdt space, and L is the space orthogonal to U.

Schölkopf et al. [118] have introduced KA into PCA. Yang et al. [120] extend Fisher discriminant analysis to associate kernel methods by incorporating KPCA and Fisher's LDA. In the LDA algorithm, firstly, the feature space is transformed by KPCA, then between-class and within-class scatter matrices S_b and S_w are computed. Finally, regular and irregular discriminant features are extracted and fused using a cumulative normalized-distance for classification.

Girolami [42] uses KPCA to develop one of the early iterative data clustering algorithms that uses KPCA. The implicit assumption of hyperspherical clusters in the sum-of-squares criterion is based on the feature space representation of the data, defined by the specific kernel chosen. Proposed generalized transform of scatter matrix is given by Equation 1.9.

$$Tr(S_W^{\theta}) = 1 - \sum_{k=1}^{K} \gamma_k \mathfrak{R}(x|C_k) \tag{1.9}$$

where $\mathfrak{R}(x|C_k)$ denotes the quadratic sum of the elements, which are allocated to the kth cluster and $\gamma_k = N_k/N$.

Liu [121] applied an extension of KPCA on Gabor-wavelet representation of face images for recognition. He uses a subset of fractional power polynomial models, which satisfies Mercer condition to compute the corresponding Gram matrix. Lemm et al. [122] used KPCA in brain imaging and makes an analytic comparison with LDA. KPCA is also used for feature space dimension reduction in biodiesel analysis [123], characterization of global germplasm [124], and bioprocess monitoring for its derivatives [125]. Zhang and Ma [126] use multiscale KPCA for fault detection in nonlinear chemical processes. Yu develops a Gaussian mixture model for fault detection and diagnosis of nonlinear chemical processes [38]. These studies are listed in Table 1.5. All these studies are shown in Fig. 1.3, with the publication year and number of citations. Table 1.5 shows that PCA approaches cover many applications, with relatively high performances.

1.2.3 Structured Database with Kernel

Structured databases pose an interesting extension of offline kernel learning. Kernel methods in general have been successful in various learning tasks on numerical (at least ordered) data represented in a single table. Much of "real-world" data, however, is structured—has no natural representation in a single table [137]. Usually, to apply kernel methods to "real-world" data, extensive preprocessing is performed to

Table 1.5 PCA approaches related to kernel algorithm

Field of application	Relevant studies	Representative accuracy
RADAR detection/prediction	[123, 128, 131]	94.6 ± 7.5
Sensor/fault detection	[125, 126, 130, 132]	96.1 ± 5.4
Prediction analysis/algorithm	[38, 41, 42, 119, 133, 135]	96.7 ± 3.4
Computer vision/biometrics	[120–122, 124, 127, 134]	99.1 ± 1.2
Geography/satellite	[129]	88.0

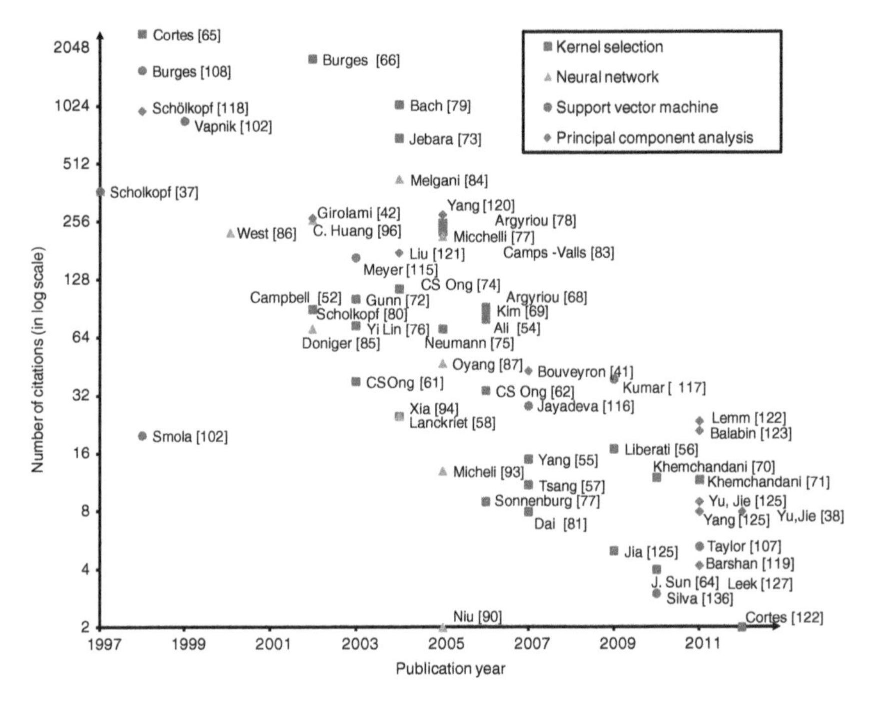

Figure 1.3 Significant research publications in various fields of KA and kernel selection for offline learning. Works related to neural network, SVM, PCA, and kernel selection are displayed according to citation number versus publishing year.

embed the data into a real vector space and thus in a single table. These preprocessing steps vary widely with the nature of the corresponding data. Several approaches of defining positive definite kernels on structured instances exist. Under this condition, structured databases are handled in different manners by analytically optimized algorithms. We briefly discuss those five kernel methods respectively: (i) kernel partial least squares (KPLS) regression [138], (ii) kernel canonical correlation analysis (KCCA) [139], (iii) kernel-based orthogonal projections (KOP), (iv) kernel multivariate analysis (KMA), and (v) diffusion kernels (DK) [140]. For the interested readers, we also include a comprehensive list of recent significant works on kernel methods for structured databases in Table 1.6.

Table 1.6 Application of kernels for structured databases

Method	Field of application	Research works
KPLS [138]	Algorithm	[141–143]
	Signal processing	[144, 145]
	Pattern recognition/Classification	[145, 146]
KCCA [139, 147, 148]	Classification	[149, 150]
	Statistical analysis	[151, 153]
KOP [166, 167]	Data compression	[152, 154]
	Classification	[155–157]
KMA [158, 159]	Image analysis/Classification	[160, 161]
	Statistical analysis	[162, 163]
DK [140]	Computer vision/Graph analysis	[164, 165]

Kernel Partial Least Squares Regression A family of regularized least squares regression models in a Reproducing Kernel Hilbert Space is extended by the KPLS regression model [138]. Similarly to principal components regression (PCR), partial least square (PLS) is a method based on the projection of input (explanatory) variables to the latent variables (components). However, in contrast to PCR, PLS creates the components by modeling the relationship between input and output variables while maintaining most of the information in the input variables. PLS is useful in situations where the number of explanatory variables exceeds the number of observations and/or a high level of multicollinearity among those variables is assumed.

Kernel Canonical Correlation Analysis (KCCA) Lai et al. propose a neural implementation of the statistical technique of canonical correlation analysis (CCA) [147] and extend it to nonlinear CCA. They derive the method of kernel-based CCA and compare these two methods on real and artificial data sets before using both on the blind separation of sources [139]. Hardoon et al. develop a general method using kernel canonical correlation analysis to learn a semantic representation to web images and their associated text. The semantic space provides a common representation and enables a comparison between the text and images [147]. Melzer et al. [148] introduce a new approach to constructing appearance models on the basis of KCCA.

Kernel-based Orthogonal Projections (KOP) Kernel-based classification and regression methods have been applied to modeling a wide variety of biological data. The kernel-based orthogonal projections to latent structures (K-OPLS) method offers unique properties facilitating separate modeling of predictive variation and structured noise in the feature space [166]. While providing prediction results similar to other kernel-based methods, K-OPLS features enhance interpretational capabilities—allowing detection of unanticipated systematic variation in the data such as instrumental drift, batch variability, or unexpected biological variation.

The orthogonal projections to latent structures (OPLS) method have been successfully applied in various chemical and biological systems as well for modeling and interpretation of linear relationships between a descriptor matrix and response matrix. A kernel-based reformulation of the original OPLS algorithm is presented where the kernel Gram matrix is used as a replacement for the descriptor matrix. This enables usage of the "kernel trick" to efficiently transform the data into a higher dimensional feature space where predictive and response-orthogonal components are calculated. This strategy has the capacity to considerably improve predictive performance, where strong nonlinear relationships exist between descriptor and response variables with retaining the OPLS model framework. The algorithm enables separate modeling of predictive and response-orthogonal variation in the feature space. This separation can be highly beneficial for model interpretation purposes, which provides a flexible framework for supervised regression [167].

Kernel Multivariate Analysis (KMA) Feature extraction and dimensionality reduction are important tasks in many fields of science dealing with signal processing and analysis. The relevance of these techniques is increasing as current sensory devices are developed with ever higher resolution, and problems involving multimodal data sources become more common. A plethora of feature extraction methods are available in the literature collectively grouped under the field of multivariate analysis (MVA) [158]. Other methods for classification and statistical dependence deal with the extreme cases of large-scale and low-sized problems. Aitchison and Aitken [159] develop an extension of the kernel method of density estimation from continuous to multivariate binary spaces. This simple nonparametric has consistent properties in discrimination problems, with some advantages over already proposed parametric counterparts.

Diffusion Kernels (DK) DK may be viewed as a distinguished example of exponential kernels [140]. These applications of DK-based learning algorithms are applicable to real valued data and a few special data types, such as strings. Kondor and Lafferty [140] propose a general method of constructing natural families of kernels over discrete structures on the basis of the matrix exponentiation idea.

1.3 DISTRIBUTED DATABASE WITH KERNEL

Various KA using offline learning may be extended to address big data by considering distributed databases. In this section, we show how KA adaptation can benefit distributed databases by the procedure described in the following three subsections: Sections 1.3.1–1.3.3.

1.3.1 Multiple Database Representation

Providing enhanced multiple databases to users has received a lot of attention in the field of data engineering [43, 44, 168–172]. As shown in Fig. 1.4, the multiple databases consist of various datasets in the forms of distributed databases and single databases. They are sometimes separately stored on multiple storage spaces in a network [169, 170, 173].

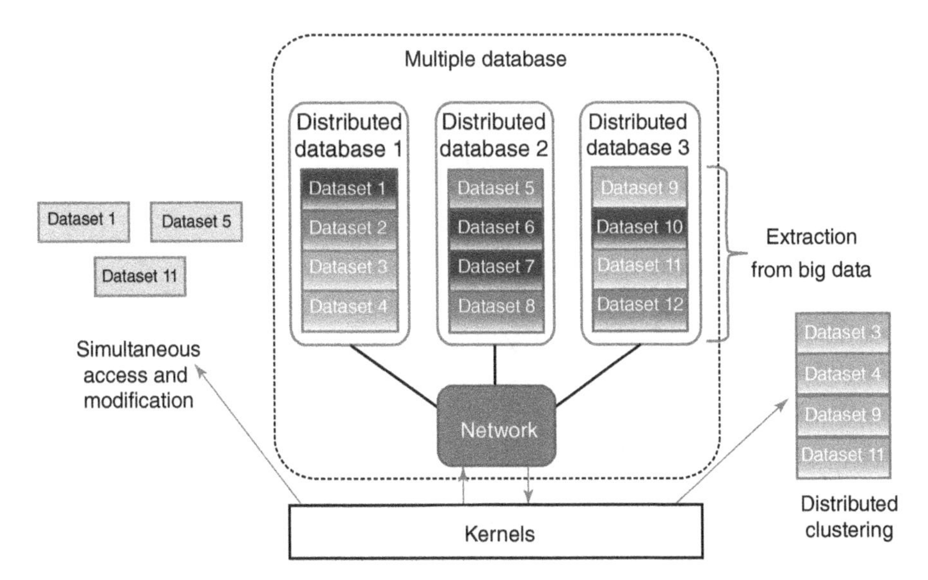

Figure 1.4 The composition of the multiple databases. The multiple databases include distributed databases and single databases, which are stored on different storage spaces connected within a network. Three key attributes of the multiple databases are (i) simultaneous access and modification, (ii) extraction from huge amounts, and (iii) distributed clustering/classification.

The multiple databases have the following three key attributes. Firstly, users of multiple databases can access and modify data in the distributed database simultaneously [169, 170]. Secondly, they can extract what they need among a huge amount of data in the distributed environments. Thirdly, all data stored in multiple databases can be clustered/classified [43, 44, 171, 172].

In order to bolster these key attributes of the multiple database, the data needs to be efficiently processed in separated storage spaces, to be classified, or to be gathered accurately per the requests of users [43, 44, 171, 172, 174]. Hence, the studies of multiple databases can be linked to multiple-learning or meta-learning, clustering, and classification problems. KA-based machine learning can be effective solutions of those issues due to KA flexibility [54, 70, 169].

1.3.2 Kernel Selections among Heterogeneous Multiple Databases

Sets of KA can be used to sort distributed databases for quick and efficient access. Lin shows that Kernel SVM can be used to decrease the total loading time of large distributed datasets at the cost of increased processing [175–178]. This is important because the data may not be accessible otherwise. The reason for the increased computational load is that the Kernel SVM must be retrained with each new dataset [175]. That is why kernels themselves are distributed and the computational power is divided. An early method of such distributed KA is described in [179]. Other more

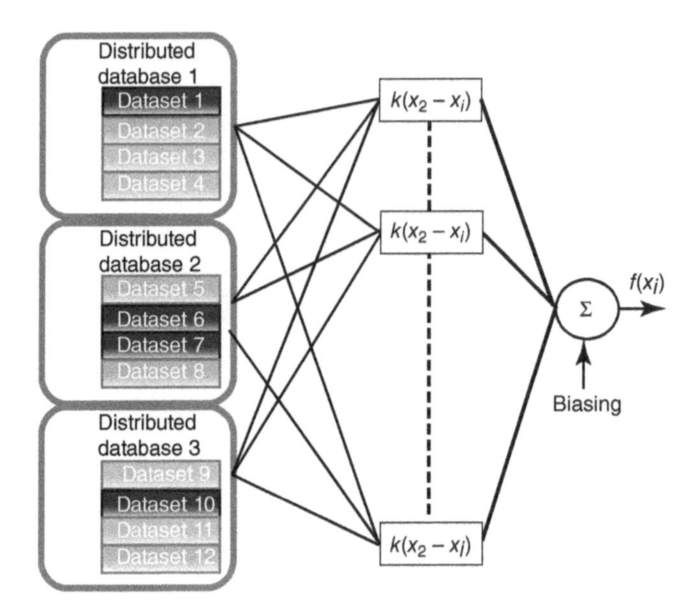

Figure 1.5 Several pregenerated kernels are trained to access different features in several distributed databases. The output of this kernel network is a sum of the kernel functions that maps to the data of interest. The summation and biasing make this method behave similarly to a neural network [178].

recent methods can be seen in [16, 178, 180–183]. Fig. 1.5 gives an example where multiple kernels (MKs) can be used in a network setting, in which each kernel holds a different configuration of the dataset [178]. The non-kernel SVM has a faster run-time but less precision than that of KA distributed SVM [175, 184].

1.3.3 Multiple Database Representation KA Applications to Distributed Databases

KA is applied for a wide range of learning problems in distributed environment. Some examples of these applications include anomaly/fault detection, medical field/biometrics/computer vision, and economic modeling/forecasting/prediction. In Table 1.7, some relevant KA applications of distributed databases are listed. Table 1.7 shows that many SVM approaches exist for a broad range of applications.

We show specific applications as follows to demonstrate how KA distributed databases using SVM-related algorithms are practically implemented in the real-world setting: (i) market analysis, (ii) wireless sensor networks, and (iii) distributed text classification.

Market Analysis One field that will benefit from the distributed computing task is the study of market behavior because it involves large and diverse datasets from

Table 1.7 Distributed database application-wise concentration

Distributed database application field	NN	SVM	PCA
Anomaly/Fault detection	[43]	[48, 171, 176, 177, 182]	[173]
	100	90.2 ± 10.7	100
Medical field/Biometrics/ Computer vision	[170]	[136, 168, 172, 180, 184, 187]	[44, 45]
	96.2	98.1 ± 1.1	91 ± 8.5
Economic modeling/ Forecasting/Prediction	[169]	[16, 174, 178, 185, 186, 188]	[239]
	95	90.0 ± 10.7	99.2

various sources. In [186], customer behavior prediction is stored in separate places, so that the proposed machine learning technique can be implemented in a distributed manner. Multiple kernel support vector machines (MKSVM) and author-developed approach called collaborative MKSVM are compared. The MKSVM is a variant of SVM that uses MKs in computation, while Collaborative MKSVM is a derivation of MKSVM to accommodate the distributed and diverse manner of data stored in a market problem. The task of learning in this study is to classify customers (returning or reluctant) and their willingness (willing to repeatedly purchase some group of products or not). Experiment results show that Collaborative MKSVM obtains higher accuracy and lower computational time and is more robust to noise data than MKSVM.

Wireless Sensor Networks This study implements SVM in classification problem over distributed WSN [16] on the basis of a sequential ascent-based algorithm. This algorithm essentially works in a parallel way with a discriminant function to appropriately handle this distributed scheme. SVM is trained locally for each network node to get the Lagrange multiplier corresponding to the local dataset. This scheme tests the data from the UCI machine learning repository [189]. Results show no significant difference in effectiveness compared to the original SVM, but the scheme minimizes the exchange of information between networks (and hence increases security) and is scalable for large-scale network.

Distributed Text Classification Another implementation of SVM is applied to classification of text in distributed computing [136]. In addition to SVM, the Relevance Vector Machine is proposed. The texts data to be classified is obtained from Reuters-21578 financial corpus and Reuters Corpus Volume 1. This text classification task is split into subtasks that can be disposed in a directed acrylic graph (DAG). The DAG is then optimized for distributed computing [190]. This distributed computing framework shows that the performance of the classification task is better using distributed scheme compared to sequential processing.

1.4 KERNEL ONLINE LEARNING

In contrast to offline learning, online learning handles one instance at a time as new data becomes available. Online learning may be used as a faster, more efficient replacement for batch processing due to its lower computational and memory requirements, or it may be implemented in a real-time system [15]. Because new data is continuously used for training, online algorithms must make use of memory resources to prevent unbounded growth of the support of the classifier or regression curve. This results in unbounded memory consumption and growth in computational complexity [15, 191–193].

In Section 1.4.1, we illustrate general principles of kernel-based online learning algorithms, and in Section 1.4.2, we demonstrate how an online kernel framework is adopted into some traditionally developed machine learning techniques such as (i) NN, (ii) SVM, and (iii) PCA. Finally, in Section 1.4.3, we establish the link between online learning and prediction.

1.4.1 Kernel-Based Online Learning Algorithms

The kernel-based online learning is introduced by shifting kernel methods to an online format in the growth of the dictionary size over time as new data is added as shown in Fig. 1.6. This growth is super-linear when batch-type offline techniques are directly applied [194] and linear when incremental methods are applied to naïve application [49]. To remedy the data size, a variety of techniques are used to discard or ignore the data on whether any new information is added to the classifier or filter [15, 49].

The so-called "growing sum problem" [192] is tackled by a variety of means. The NORMA algorithm [15] uses a "forgetting factor" such that samples are weighted in a sliding-window manner, with older samples weighted less than newer ones. This technique assumes that we are interested only in the statistical structure of the newest samples in a nonstationary series.

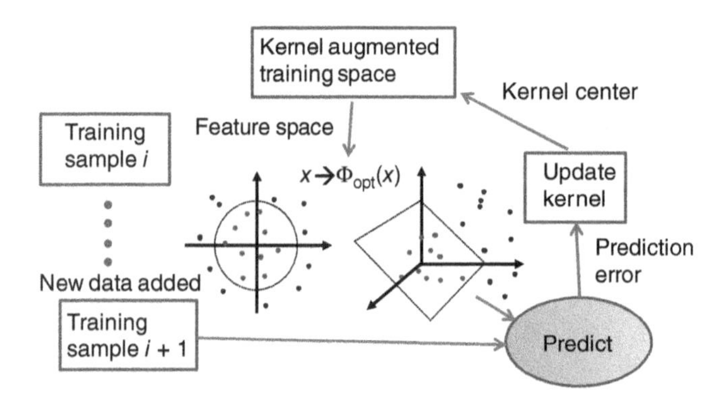

Figure 1.6 Online kernel-based learning algorithm. As a new training sample is added, the kernel is updated on the basis of the online learning algorithms described in the following section.

Alternatively, when it is desired to take older data into account, sparsification techniques may be used [49]. In these methods, only data that add new information to the classifier or filter are added to the support. Various information criteria are proposed, such as the approximate linear dependency criterion [195], the surprise criterion [50], and the variance criterion [196].

While sparsification methods seek to discard redundant data, quantization methods instead limit the number of possible input space regions that will lead to support expansion [197]. The input space is divided into discrete regions of a given size. The region expands the support only if it has no assigned support vector. Otherwise, the coefficient for that region is updated instead, which limits the growth of the system [197].

Other methods are proposed for dealing with the growth problem, including [192], which uses only finite-dimensional kernels and replaces infinite-dimensional kernels with finite-dimensional approximations. This method achieves constant computational complexity over time [192]. Another method is presented in [193], which projects new inputs into a feature space spanned by the previous hypothesis. This method is guaranteed to be bounded and outperforms other Perceptron-based methods [193].

The complication of online kernel methods is the selection of the kernel itself. Often, it is assumed that a suitable function is known as *a priori*, while in reality, this is not necessarily the case [49, 198]. This issue is less well addressed in the literature but is handled through the use of MK algorithms, which select the best kernel or combination of kernels from a dictionary of possible choices on the basis of the data [198, 199]. Rosenblatt [200] uses a combination of the Perceptron algorithm and the Hedge algorithm [191] to learn the kernel combination and the classifier and present both deterministic and stochastic [201] algorithms for doing so.

1.4.2 Adopt "Online" KA Framework into the Traditionally Developed Machine Learning Techniques

Figure 1.7 shows the representative online learning studies on Kernel-based online learning using (i) NN, (ii) SVM, and (iii) PCA.

Neural Network (NN) Online NN is designed to have the learning capabilities of online time systems [202]. Online NN also consists of input, hidden, and output layers interconnected with directed weights (w). w_{ij} denotes the input-to-hidden layer weights at the hidden neuron j, and w_{jk} denotes the hidden-to-output layer weights at the output neuron k. Learning is accomplished by comparing the actual output to the desired output, then adjusting the weights accordingly. A typical kernel online NN is depicted in Fig. 1.8.

Online updates to the weights of a NN relies on algorithms such as the commonly used Gaussian RBF, which resides in the hidden layer of the NN. The centers of the RBF function are initially chosen on the basis of prior knowledge of the structure of the uncertainty, and typically, we assume that the RBF centers are fixed. The application of kernel methods allows this assumption to be relaxed [219].

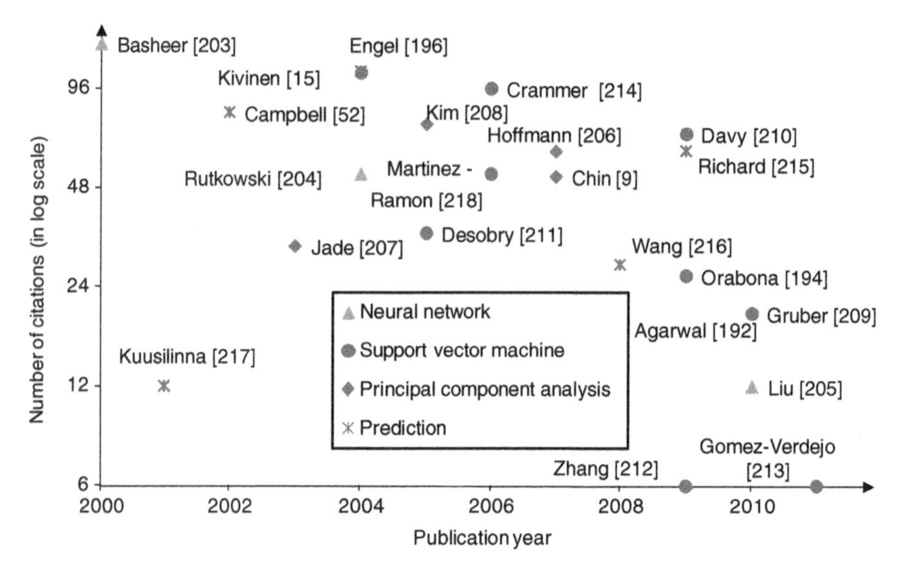

Figure 1.7 Key papers for Kernel Online learning based on the number of citations for the years 1998–2013, for Principal Component Analysis (PCA), Support Vector Machine (SVM), Neural Network (NN), Autoregressive Moving Average (ARMA), and Finite-State Model (FSM).

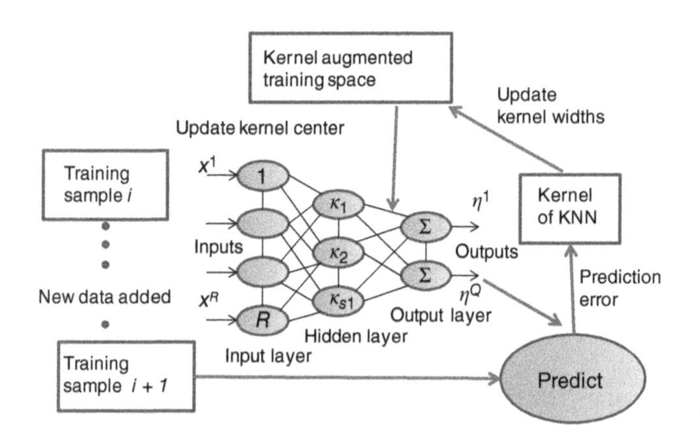

Figure 1.8 Online NN learning algorithm. An artificial neural network consists of input, hidden, and output layers interconnected with directed weights (w), where w_{ij} denotes the input-to-hidden layer weights at the hidden neuron j and w_{jk} denotes the hidden-to-output layer weights at the output neuron k [218].

Kingravi et al. propose an online algorithm that connects kernel methods to persistently exciting (PE) signals, known as budgeted kernel restructuring (BKR) for model reference adaptive control (MRAC).

Similar techniques have been applied to such problems as time-series prediction [203], and pattern classification tasks such as facial recognition [15]. NN is frequently applied in engineering applications that require optimization of nonlinear stochastic dynamic systems [220]. Yuen et al. [218] propose a kernel-based hybrid NN for online regression of noisy data, combining fuzzy ART (FA) and general regression neural network (GRNN) models. The performance of kernel NN continues to improve due to the recent development of new algorithms; Liu et al. [204] describe kernel affine projection algorithms (KAPA), which outperforms other recently developed algorithms such as the kernel least-mean-square (KLMS) algorithm in online time-series prediction, nonlinear channel equalization, and nonlinear noise cancellation.

Support Vector Machine (SVM) Online SVM [208, 209] preserves the skeleton samples on the basis of the geometric characteristics of SVM as shown in Fig. 1.9. The SVM classifier is updated when the distances between samples in the kernel space and the newly arriving samples are within a given threshold to the current classification hyperplane. Online updating of the classifier can be achieved, as only very limited training samples are maintained in the offline step [208].

The implementation of online SVM has two possible cases. One case is to replace batch processing when more than a feasible amount of memory is required to generate the classifier. The second case involves data arriving in real time, when the algorithm classifies data as it is received. In this case, we consider nonstationary aspects so that the classifier can adapt—that is, retrain—to the changing distribution of the input data. This illuminates the need for specially designed algorithms, as the training process for traditional batch SVMs is notoriously slow [15].

Gruber et al. adopt an online SVM to signature variation, using kernel functions based on the longest common subsequences (LCSS) detection algorithm. The model compensates local variability of individual signatures, performs more reliably than other SVM kernel algorithms such as dynamic time warping, and easily defeats them [208].

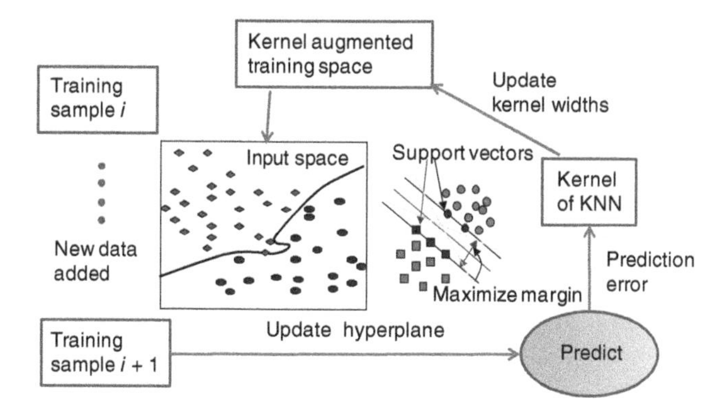

Figure 1.9 Scheme for naïve online SVM classification. The input data are mapped to the kernel Hilbert space, and the hyperplane is generated as in offline SVM. If an optimal solution cannot be found, a new kernel is selected. The process repeats for newly incoming data.

Online SVM is also widely used for novelty detection, known as anomaly detection. The novelty detection is the process when a learning system recognizes unprecedented inputs [209]. SVMs adapted to online novelty detection are referred to as one-class support vector machines (1-SVM). 1-SVM distinguishes between inputs in the "normal" class and all other inputs, which are considered outliers or anomalies. Because such a classification does not require expert labels, 1-SVMs are considered unsupervised methods. The classifying function of a 1-SVM returns +1 within a region capturing most of the data points, and −1 elsewhere in the input space [204]. This is accomplished by separating the data from the origin using the greatest possible margin [204, 221]. Similarly to other online SVMs, online 1-SVMs typically prune old or redundant data from the classifier to prevent indefinite complexity expansion [209].

Desobry et al. [210] propose an adaptive online 1-SVM method called kernel change detection (KCD), designed for the online case of the sequential input data. For T feature vectors $\boldsymbol{x}_t$, $t = 1, 2, \ldots T$, they train two 1-SVMs independently: one on the set of preceding feature vectors, and one on the set of future vectors. Using the outputs from each SVM, they compute a decision index and compare it to a predetermined threshold to ascertain whether a change has occurred in the feature statistics at the given t. They find that KCD is capable of segmenting musical tracks into their component notes.

Zhang et al. implement a sliding-time-window 1-SVM using quarter-sphere formulation of the SVM, which applies to one-sided non-negative distributions of features [211]. Rather than fitting a hyperplane to the data, this method fits a quarter-(hyper)sphere, centered at the origin [211]. This reduces the quadratic programming solution of the SVM dual problem to a linear programming solution, reducing the computational effort required for training at each time window.

Gomez-Verdejo et al. [212] create an adaptive one-class support vector achine (AOSVM) that updates the SVM at every time-step using a recursive least squares algorithm. AOSVM weights old patterns with an exponential window and allows them to decay so that the SVM can adapt to changing data statistics. Sofman et al. [222] use a modified version of the NORMA algorithm [15] for online novelty detection in the context of mobile robot terrain navigation. NORMA is a stochastic gradient descent algorithm suitable for online kernel-based learning. Sofman uses a hinge loss function, such that its gradient is nonzero only in the case of novel inputs.

Principal Component Analysis (PCA) Online KPCA extracts principal components in the feature space related to nonlinear inputs. The algorithm for extracting the components is Equation 1.10 expressed mathematically as

$$V^n \cdot \Phi(x) = \sum_{i=1}^{M} \alpha_i^n k(x_i, x) \tag{1.10}$$

where $k(x_i, x) = (\Phi(x_i) \cdot \Phi(x))$ is the kernel function, α_i the eigenvectors, and V^n the corresponding eigenvectors in the feature space [118]. The success of online KPCA depends heavily on the kernel updates from the modified Gram matrix; thus, the classification accuracy relies on kernel iteration for problem-specific datasets [240–242].

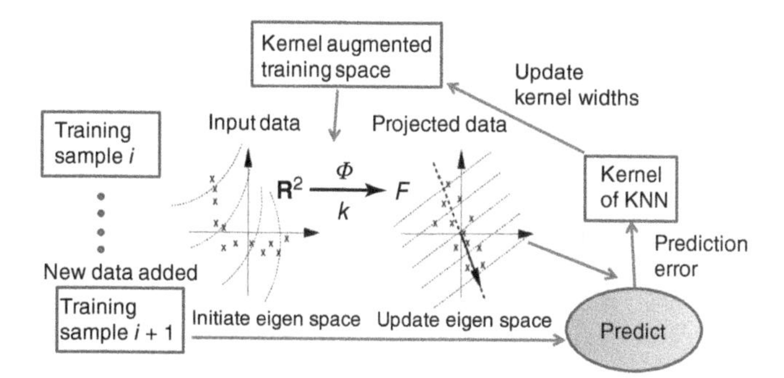

Figure 1.10 Online KPCA diagram. In the initial (offline) training phase, the initial kernel parameters and eigen-feature space are calculated. As new training sets are introduced, incremental (online) adjustments are made to the parameters and an updated eigen-space is generated.

Most online methods begin with one or more offline learning steps to determine suitable kernel parameters [223]. The incremental kernel principal component analysis (IKPCA) developed by Ozawa et al. [224] progresses in two stages: (i) in the initial learning phase, a number of training sets are presented offline and the initial eigen-feature space is found, and (ii) in the incremental learning phase, the feature space is defined by solving the eigenvalue problem with dimensions determined by the amount of independent data. Online KPCA algorithm is depicted in Fig. 1.10.

Hoffman [205] demonstrates the use of online KPCA in novelty detection, specifically applied to recognition of handwritten digits and breast cancer detection. Classification performance is measured against results from the Parzen window density estimator, standard PCA, and 1-SVM. By adjusting the parameters q (number of eigenvectors) and σ (kernel width), KPCA achieves better generalization than Parzen density and greater resistance to noise. Online KPCA methods are applied to the denoising of chaotic time series [206] and Gaussian noise in images [207].

1.4.3 Relationship Between Online Learning and Prediction Techniques

Prediction is a scheme to estimate the future behavior of a system. It is intimately connected with regression [52, 225] and system modeling [51, 214], each of which seeks to build a mathematical representation of an unknown system or process. This mathematical representation model can be subsequently used to predict future states.

Prediction and online learning have traditionally been studies of two independent communities: (i) in signal processing and (ii) in machine learning. The two disciplines possess different momentum and emphasis, which makes it attractive to periodically review trends and new developments in their overlapping spheres of influence. These two existing communities notice that this is the key area of study for the next generation of any intelligent systems [191, 214, 215].

Table 1.8 Comparisons of prediction and online learning algorithms

Methods	Purpose	Main community
Prediction	Estimate forthcoming outcomes	Statistics/signal processing
Online learning	Induce the coming new data for prediction	Statistics/machine leaning

As shown in Table 1.8, the term "prediction" implies learning how to forecast, that is, what will happen in the forthcoming data? Online learning, on the other hand, is a more general term, in which newly arriving data may be used to update a classifier without necessarily generating a prediction of the next data point. However, they are often used to mean the same thing, as knowledge of the structure of a system allows one to predict its future behavior [52, 225].

A prediction is a future function, as a forecast of the new coming dataset will be automatically classified on the basis of the knowledge of offline or online learning. As shown in the previous section, there are many prediction approaches. These predictions can be incorporated with KA representation, in a manner that may extend the learning space through the nonlinear online learning. In the following section, "prediction" will be examined, specifically how the coming datasets are used for testing a future feature space prediction [213, 214] as "online learning" has been shown.

1.5 PREDICTION WITH KERNELS

Prediction is represented by top-down model-based representation. In this section, we show how kernels may enhance the prediction through the four major representations: (i) linear prediction, (ii) Kalman filter, (iii) finite-state model, and (iv) autoregressive moving average model. Finally, we compare them in (v) comparison of four models.

1.5.1 Linear Prediction

A linear prediction is a simplified model where future output values are estimated as a linear function (Eq. 1.11) of previous values and predictor coefficients, as follows [226]:

$$\hat{x}(n) = a_0 + a_1 x(n-1) + \cdots + a_n x(n-k) = \sum_{i=0}^{k} a_i x(n-i) \tag{1.11}$$

where $\hat{x}\,(n)$ is the predicted value or position at n. By applying a kernel to the input data, the dataset can be mapped to a space where it is linear.

The predicted value is a linear combination of previous observations $x(n-k)$ and predictor coefficients a_n, as shown in Fig. 1.11. The key is to solve a linear equation to find out the coefficients, a_n, that can minimize the mean squared error between the predicted values and previous values [227]. The linear model is widely used in the early stage to compare the prediction performance with other models, for example, Kalman filtering [226].

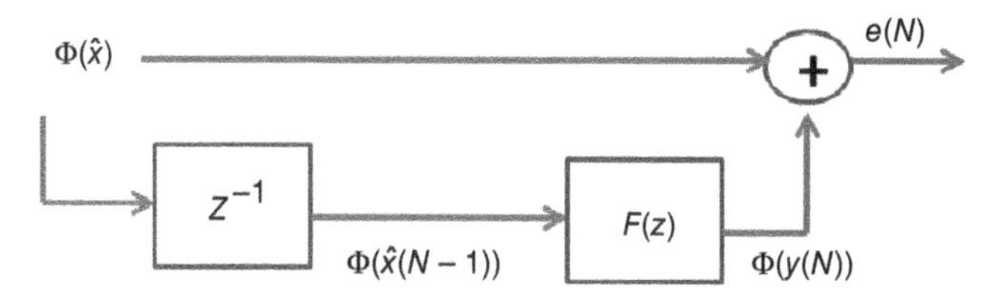

Figure 1.11 Linear predictor with an initial kernel mapping. The input data is mapped from its original space to a new Hilbert space where $x(n)$ is linear, by the function Φ to allow for the use of the kernel trick. The model is optimized by minimized the function $e(N)$, which is the difference in the current and previous predicted value.

1.5.2 Kalman Filter

The Kalman filter (KF) is one of the most commonly used prediction methods in real-time filtering technologies [53, 226, 228–233]. KF provides a recursive solution 1.12 to minimize mean square error within the class of linear estimators, where linear process and measurement equations can be expressed as follows [228]:

$$\hat{x}(t) = Fx(t-1) + Bu(t-1) + W, \ \ z(t) = H\hat{x}(t) + V \tag{1.12}$$

where we denote the state transition matrix as F, the control-input matrix as B, and the measurement matrix as H. $u(t)$ is an n-dimensional known vector, and $z(t)$ is a measurement vector. The random variables W and V represent the process and measurement noise with the property of the zero-mean white Gaussian noise with covariance, $E[W(t)W(t)^T] = R(t)$ and $E[V(t)V(t)^T] = Q(t)$, respectively. The matrices F, B, W, H, and V are assumed known and possibly time varying.

Ralaivola and d' Alche-Buc [233] develop a kernel Kalman filter (KKF) method for implementing a KF on a nonlinear time-series dataset. By mapping the input data from a nonlinear space to a Hilbert space (shown in Fig. 1.12), the mercer kernel function can be satisfied. In KKF, the predicted position $\hat{x}(t)$ can be derived from the previous state $x(t-1)$ and the current measurement $z(t)$ [226, 232]. Because of the state update kernel process with new data, KF is effective for predicting changes in both linear and nonlinear dynamic systems of multivariate Gaussian distribution [53].

KF can be enhanced to an interactive multiple model (IMM) filter with constant velocity (CV) and constant acceleration (CA) [232]. Hong et al. [53] suggest the first-order extended Kalman filter (EKF) with a kernel type method can be used to process and update the state estimate.

1.5.3 Finite-State Model

Finite-state model (FSM), or finite-state machine, consists of a finite number of discrete states, only one of which the model can occupy at any given time. FSM undergoes a change of state upon a triggering event. Kernel methods have been applied to FSM [234, 235]. Separated-kernel image processing using finite-state machines (SKIPSM) is an online kernel method used to speed processing time when dealing

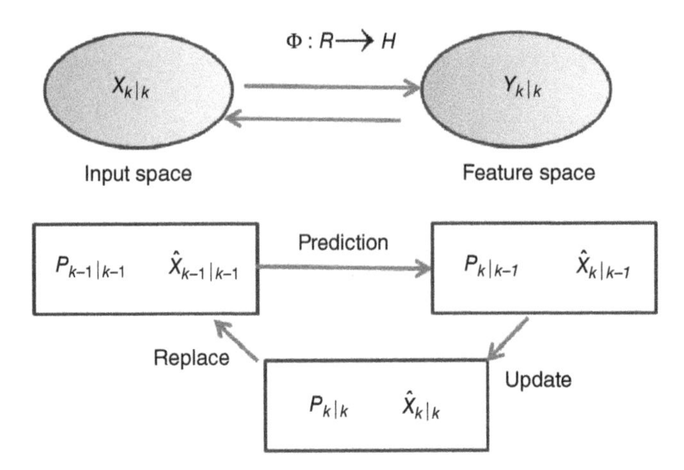

Figure 1.12 For processing a nonlinear time series, a kernelized Kalman filter is an optimal way to execute the Kalman filter. With a nonlinear series x_k and an associated series $x^{\Phi}{}_k$. x_k is mapped to $x^{\Phi}{}_k$ by the nonlinear map $\Phi: \ R^d \rightarrow H$, where R^d is the input space, H is the feature space, and $[\Phi(x), \ \ \Phi(y)] = k(x, y)$.

with large numbers of states, as in processing of grayscale images [234]. FSM is expressed mathematically as a sextuple $M = (I, \ O, \ S, \ \delta, \ \lambda, \ s0)$, where I is the input alphabet, O is the output alphabet, S is the finite set of states, and $s0$ is the initial state. δ is the next state function ($\delta : S \times I \rightarrow S$), and λ is the output function ($\lambda : S \times I \rightarrow O$). Most FSM are classified as either a Mealy machine, in which the inputs always affect the outputs, or a Moore machine, in which inputs affect outputs only through the states [216, 236].

1.5.4 Autoregressive Moving Average Model

Autoregressive moving average (ARMA) model is a mathematical generalization of the linear model with time-series data and signal noise and is widely used to predict motion patterns of a time series from past values. Martinez-Ramòn et al. [217] use the kernel trick to formulate nonlinear SVM-ARMA in reproducing kernel Hilbert space (RKHS). Discrete-time processes (DTP) that are nonlinear cannot be directly modeled using ARMA, but the use of composite kernels allows for a nonlinear mapping into an RKHS. The input and output DTP are represented by the composite kernel formulation 1.13:

$$\begin{aligned} K(z_i, z_j) &= F\langle[\phi_y(y_{i-1})^T, \phi_u(u_i)^T]^T, [\phi_y(y_{j-1})^T, \phi_u(u_j)^T]^T\rangle \\ &= K_y(y_{i-1}, y_{j-1}) + K_u(u_i, u_j) \end{aligned} \tag{1.13}$$

where $\boldsymbol{y}$ and $\boldsymbol{u}$ are the input and output DTP, respectively. Shpigelman et al. [237] describe a kernel autoregressive moving average (KARMA) for the tracking of hand movement in 3D space.

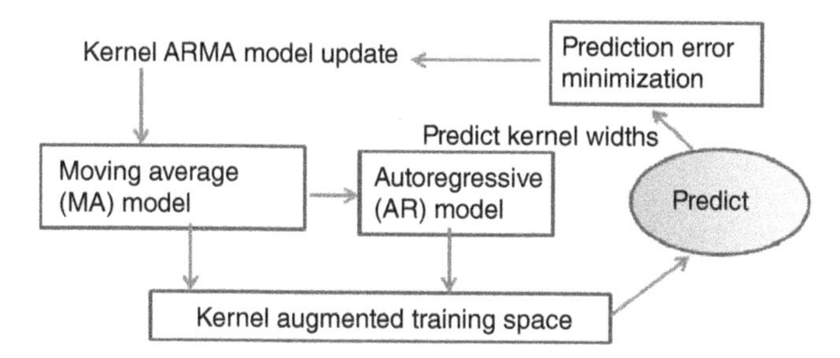

Figure 1.13 A kernel autoregressive moving average (KARMA) model with online learning for tracking of hand movement.

As shown in Fig. 1.13, ARMA consists of two models: (i) an autoregressive (AR) model represented by a weighted sum of the present and past positions with a polynomial order p, and (ii) a moving average (MA) model represented by a weighted sum of the present and past signal noise with a polynomial order q. The notation ARMA(p, q) is represented by the polynomial orders of p AR and q MA [237, 238].

1.5.5 Comparison of Four Models

The four prediction models, linear prediction, KF, FSM, and ARMA, are quantitively compared to illustrate the overall advantages of each model, as shown in Table 1.9. The prediction error is defined as root mean square error (RMSE $= \sqrt{(\Sigma(y_i - \hat{y}_i)^2 / \Sigma(y_i - m_y)^2)}$, where y_i is the ith measurement, $\hat{y}_i$ is the estimation of the ith measurement, and m_y is the mean of all the measurements). RMSE is used to convert each quantitive error to a universal metric for proper comparison. On the basis of these various conditions, we cannot definitively state which method performs better than the others. The performance of the computational speed is separately examined for the comparison by listing the primary factor contributing to model complexity.

These methods also list in Table 1.10 the computational speed to describe the overall computational complexity. In general, the number of datasets dictates the speed of each method, although some methods depend on additional experimental factors. For example, the main factors to determine the computational speed using Kernel Predictive Linear Gaussian models [227] are the number of base functions and total amount of feature dimensions.

Table 1.9 Quantitative comparisons of prediction error

Methods	Error range	Mean ± SDV	Data type
Linear [227]	[0.0,0.05]	0.026 ± 0.018	k-fold cross-validation
KF [233]	[0.03, 0.73]	10.37 ± 10.62	Three noisy datasets
FSM [234, 235]	[0.0, 0.135]	0.042 ± 0.042	10 uniform
ARMA [217]	[−2.9, −1.7]	-2.3 ± 0.60	Normalized validation

Table 1.10 Comparisons of computational speed

Methods	Speed	Main variables
Linear [227]	$O(Jn^3)$	J: base functions n: feature dimension
KF [233]	$O(l^3)$	l: time series length
FSM [234, 235]	8.34 ± 7.2 frame/s	Morphological structure
ARMA [217]	$O(n)$	n: data

1.6 FUTURE DIRECTION AND CONCLUSION

In this survey, we have shown the state-of-the-art techniques related to KA, mainly for classification and prediction including online and distributed extension and selected applications. We have also explained KA prediction approaches including model-based and hybrid prediction algorithms. The cross-section between classification and prediction is a new research area, considering the uncertainty and irregularity of data sequential configuration. Originating from physical phenomena and technical limitations, these open areas may lead to further investigations of sophisticated KA techniques. In successful implementations of KA for big-data analysis, the prediction and online classification are sometimes merged together. Unlike classical partitioning clustering algorithms for offline machine learning, modern KA for online learning and prediction are able to solve real-world problems despite the big data size. To maximize the potential innovation furthermore, an extensive validation on real-world applications remains a big challenge for these KA techniques. To realize this future direction, collaborative research activities with various disciplines including engineering (signal processing), computer science (machine learning), and statistics (data regression) are preferable.

REFERENCES

1. T. Hoya, and Y. Washizawa, “Simultaneous pattern classification and multidomain association using self-structuring kernel memory networks,” *IEEE Trans. Neural Netw.*, vol. 18, no. 3, pp. 732–744, 2007.
2. B. J. Kim, I. K. Kim, and K. B. Kim, “Feature extraction and classification system for nonlinear and online data,” in *Proceedings of Advances in Knowledge Discovery and Data Mining*, Sydney, Australia, May 26–28, vol. 3056, pp. 171–180, 2004.
3. W. Zheng, C. Zou, and L. Zhao, “An improved algorithm for kernel principal component analysis,” *Neural Process. Lett.*, vol. 22, no. 1, pp. 49–56, 2005.
4. V. N. Vapnik, *The nature of statistical learning theory*, 2nd ed. New York: Springer, 2000.
5. R. Duda, P. Hart, and D. Stock, *Pattern Classification*, 2nd ed. Hoboken, NJ: John Wiley & Sons, Inc, 2001.
6. T. Briggs and T. Oates, “Discovering domain specific composite kernels,” in *Proc. 20th Nat'l. Conf. Artificial Intelligence*, Pittsburgh, Pennsylvania, July 9–13, pp. 732–738, 2005.
7. N. Cristianini, J. Kandola, A. Elisseeff, and J. Shawe-Taylor, “On kernel target alignment,” in *Proc. Neural Information Processing Systems*, British Columbia, Canada, Dec. 3–8, pp. 367–373, 2001.

8. H. Xiong, Y. Zhang, and X. W. Chen, "Data-dependent kernel machines for microarray data classification," *IEEE/ACM Trans. Comput. Biol. Bioinform.*, vol. 4, no. 8, pp. 583–595, 2007.
9. T. J. Chin and D. Suter, "Incremental kernel principal component analysis," *IEEE Trans. Image Process.*, vol. 16, no. 6, pp. 1662–1674, 2007.
10. L. Winter, Y. Motai, and A. Docef, "Computer-aided detection of polyps in CT colonography: On-line versus off-line accelerated kernel feature analysis," Special Issue on Processing and Analysis of High-Dimensional Masses of Image and Signal Data, *Signal Process.*, vol. 90, pp. 2456–2467, 2010.
11. Y. Motai, "Synthesized articulated behavior using space-temporal on-line principal component analysis," Special Issue on Imitative Robots, *Adv. Robotics*, vol. 21, no. 13, pp. 1503–1520, 2007.
12. M. Awad, Y. Motai, J. Nappi, and H. Yoshida, "A clinical decision support framework for incremental polyps classification in virtual colonoscopy," *Algorithms*, vol. 3, pp. 1–20, 2010.
13. S. Li and J. Lu, "Face recognition using the nearest feature line method," *IEEE Trans. Neural Netw.*, vol. 10, no. 2, pp. 439–443, 1999.
14. C. Park and S. B. Cho, "Genetic search for optimal ensemble of feature-classifier pairs in DNA gene expression profiles," in *Proc. Int. Joint Conf. Neural Networks*, Portland, Oregon, July 20–24, vol. 3, pp. 1702–1707, 2003.
15. J. Kivinen, A. J. Smola, and R. C. Williamson, "Online learning with kernels," *IEEE Trans. Signal Process.*, vol. 52, no. 8, pp. 2165–2176, 2004.
16. D. Wang, J. Zheng, Y. Zhou, and J. Li, "A scalable support vector machine for distributed classification in ad hoc sensor networks," *Neurocomputing*, vol. 75, no. 1–3, pp. 394–400, 2010.
17. X. Jiang, R. Snapp, Y. Motai, and X. Zhu, "Accelerated kernel feature analysis," in *Proc. IEEE Computer Society Conf. Computer Vision and Pattern Recognition*, New York, New York, June 17–22, pp. 109–116, 2006.
18. T. Damoulas and M. A. Girolami, "Probabilistic multi-class multi-kernel learning: On protein fold recognition and remote homology detection," *Bioinformatics*, vol. 24, no. 10, pp. 1264–1270, 2008.
19. H. Xiong, M.N.S. Swamy, and M.O. Ahmad, "Optimizing the data-dependent kernel in the empirical feature space," *IEEE Trans. Neural Netw.*, vol. 16, pp. 460–474, 2005.
20. Y. Motai and H. Yoshida, "Principal composite kernel feature analysis," *IEEE Trans. Knowl. Data Eng.*, vol. 25, no. 8, pp. 1863–1875, 2013.
21. S. Ozawa, S. Pang, and N. Kasabov, "Incremental learning of chunk data for online pattern classification systems," *IEEE Trans. Neural Netw.*, vol. 19, no. 6, pp. 1061–1074, 2008.
22. H. T. Zhao, P. C. Yuen, and J. T. Kwok, "A novel incremental principal component analysis and its application for face recognition," *IEEE Trans. Syst. Man Cyb. B*, vol. 36, no. 4, pp. 873–886, 2006.
23. Y. M. Li, "On incremental and robust subspace learning," *Pattern Recogn.*, vol. 37, no. 7, pp. 1509–1518, 2004.
24. Y. Kim, "Incremental principal component analysis for image processing," *Opt. Lett.*, vol. 32, no. 1, pp. 32–34, 2007.

25. B. J. Kim and I. K. Kim, "Incremental nonlinear PCA for classification," in *Proc. Knowledge Discovery in Databases (PKDD 2004)*, Pisa, Italy, Sept. 20–24, vol. 3202, pp. 291–300, 2004.
26. B. J. Kim, J. Y. Shim, C. H. Hwang, I. K. Kim, and J. H. Song, "Incremental feature extraction based on empirical kernel map," *Found. Intell. Syst.*, Maebashi City, Japan, October 28–31, vol. 2871, pp. 440–444, 2003.
27. L. Hoegaerts, L. De Lathauwer, I. Goethals, J. A. K. Suykens, J. Vandewalle, and B. De Moor, "Efficiently updating and tracking the dominant kernel principal components," *Neural Netw.*, vol. 20, no. 2, pp. 220–229, 2007.
28. B. Schölkopf and A. J. Smola, "Learning with kernels: Support vector machines, regularization, optimization, and beyond," *Adaptive computation and machine learning*. Cambridge, MA: MIT Press, 2002.
29. H. Fröhlich, O. Chapelle, and B. Schölkopf, "Feature selection for support vector machines by means of genetic algorithm," in *Proc. 15th. IEEE Int. Conf. Tools with Artificial Intelligence*, Sacramento, California, Nov. 3–5, pp. 142–148, 2003.
30. X. W. Chen, "Gene selection for cancer classification using bootstrapped genetic algorithms and support vector machines," in *Proc. IEEE Int. Conf. Computational Systems, Bioinformatics*, Palo Alto, CA, August 11–14, pp. 504–505, 2003.
31. F. A. Sadjadi, "Polarimetric radar target classification using support vector machines," *Opt. Eng.*, vol. 47, no. 4, pp. 1314–1325, 2008.
32. M. Awad and Y. Motai, "Dynamic classification for video stream using support vector machine," Special Issue on Soft Computing for Dynamic Data Mining, *J. Appl. Soft Comput.*, vol. 8, no. 4, pp. 1314–1325, 2008.
33. S. Amari and S. Wu, "Improving support vector machine classifiers by modifying kernel functions," *Neural Netw.*, vol.6, pp. 783–789, 1999.
34. B. Souza and A. de Carvalho, "Gene selection based on multi-class support vector machines and genetic algorithms," *Mol. Res.*, vol. 4, no.3, pp. 599–607, 2005.
35. R. Kondor and T. Jebara, "Gaussian and Wishart hyperkernels," in *Advances in Neural Information Processing Systems*, Vancouver and Whistler, Cananda, Dec. 4–7, vol. 19, pp. 729–736, 2006.
36. J. B. Yin, T. Li, and H.-B. Shen, "Gaussian kernel optimization: Complex problem and a simple solution," *Neurocomputing*, vol. 74, no. 18, pp. 3816–3822, 2011.
37. B. Scholkopf, K. K. Sung, C. Burges, F. Girosi, P. Niyogi, T. Poggio, and V. Vapnik, "Comparing support vector machines with Gaussian kernels to radial basis function classifiers," *IEEE Trans. Signal Process.*, vol. 45, no. 11, pp. 2758–2765, 1997.
38. J. Yu, "A nonlinear kernel Gaussian mixture model based inferential monitoring approach for fault detection and diagnosis of chemical processes," *Chem. Eng. Sci.*, vol. 68, no. 1, pp. 506–519, 2012.
39. K. L. Chan, P. Xue, and L. Zhou, "A kernel-induced space selection approach to model selection in KLDA," *IEEE Trans. Neural Netw.*, vol. 19, no. 12, pp. 2116–2131, 2008.
40. W. S. Zheng, J. H. Lai, and P. C. Yuen, "Penalized preimage learning in kernel principal component analysis," *IEEE Trans. Neural Netw.*, vol. 21, no. 4, pp. 551–570, 2010.
41. C. Bouveyron, S. Girard, and C. Schmid, "High-dimensional data clustering," *Comput. Stat. Data Anal.*, vol. 52, no. 1, pp. 502–519, 2007.
42. S. Girolami "Mercer kernel-based clustering in feature space," *IEEE Trans. Neural Netw.*, vol. 13, pp. 780–784, 2002.

43. B. Ayhan, M.-Y. Chow, and M.-H. Song, "Multiple discriminant analysis and neural-network-based monolith and partition fault-detection schemes for broken rotor bar in induction motors," *IEEE Trans. Indus. Electron.*, vol. 53, no. 4, pp. 1298–1308, 2006.

44. F. Ratle, C. Gagnéb, A.-L. Terrettaz-Zuffereyc, M. Kanevskia, P. Esseivac, and O. Ribauxc, "Advanced clustering methods for mining chemical databases in forensic science," *Chemom. Intell. Lab. Syst.*, vol. 90, no. 2, pp. 123–131, 2008.

45. P. Phoungphol and Z. Yanqing, "Multi-source kernel k-means for clustering heterogeneous biomedical data," in *Proc. IEEE Intl. Conf. Bioinformatics and Biomedicine Workshops*, Atlanta, Georgia, Nov. 12–15, pp. 223–228, 2011.

46. K. Buza, P. B. Kis, and A. Buza, "Graph-based clustering based on cutting sets," in *Proc. 15th IEEE Intl. Conf. Intelligent Engineering Systems*, Poprad, Slovakia, June 23–25, pp. 143–149, 2011.

47. G Tsoumakas, L Angelis, and I Vlahavas, "Clustering classifiers for knowledge discovery from physically distributed databases," *Data Knowl. Eng.*, vol. 49, no. 3, pp. 223–242, June 2004.

48. P. Honeine, C. Richard, J. C. M. Bermudez, and H. Snoussi, "Distributed prediction of time series data with kernels and adaptive filtering techniques in sensor networks," in *Proc. 42nd Conf. Signals, Systems and Computers*, Pacific Grove, California, Nov. 6–9, pp. 246–250, 2011.

49. M. Yukawa, "Multikernel adaptive filtering," *IEEE Trans. Signal Processing*, vol. 60, no. 9, pp. 4672–4682, 2012.

50. W. Liu, I. Park, and J. C. Principe, "An information theoretic approach of designing sparse kernel adaptive filters," *IEEE Trans. Neural Netw.*, vol. 20, no. 12, pp. 1950–1961, 2009.

51. G. Li, C. Wen, Z. G. Li, A. Zhang, F. Yang, and K. Mao, "Model-based online learning with kernels," *IEEE Trans. Neural Netw. Learn. Syst.*, vol. 24, no. 3, pp. 356–369, 2013.

52. C Campbell, "Kernel methods: A survey of current techniques," *Neurocomputing*, vol. 48, no. 1–4, pp. 63–84, 2002.

53. S.-M. Hong, B.-H. Jung, and D. Ruan, "Real-time prediction of respiratory motion based on a local dynamic model in an augmented space," *Phys. Med. Biol.*, vol. 56, no. 6, pp. 1775–1789, 2011.

54. S. Ali and K. A. Smith-Miles, "A meta-learning approach to automatic kernel selection for support vector machines," *Neurocomputing*, vol. 70, no. 1–3, pp. 173–186, 2006.

55. S. Yan, C. Zhang, and X. Tang, "Bilinear analysis for kernel selection and nonlinear feature extraction," *IEEE Trans. Neural Netw.*, vol. 18, no. 5, pp. 1442–1452, 2007.

56. C. Liberati, et al., "Data adaptive simultaneous parameter and kernel selection in kernel discriminant analysis (KDA) using information complexity," *J. Pattern Recogn. Res.*, vol. 4, no. 4, pp. 119–132, 2009.

57. I. W. Tsang and J. T. Kwok, "Efficient hyperkernel learning using second-order cone programming," *IEEE Trans. Neural Netw.*, vol.17, no. 1, pp. 48–58, 2006.

58. G. R. G. Lanckriet et al., "Learning the kernel matrix with semi-definite programming," *J. Mach. Learn. Res.*, vol. 5, pp. 27–72, 2004.

59. C. Ke and S. Zhong, "Kernel target alignment for feature kernel selection in universal steganographic detection based on multiple kernel SVM," in *Proc. Int. Symp. Instrumentation Measurement, Sensor Network and Automation*, Sanya, pp. 222–227, August 2012.

60. T. Jebara, "Multitask sparsity via maximum entropy discrimination," *J. Mach. Learn. Res.*, vol. 2, pp. 75–110, 2011.
61. C. S. Ong, J. Smola, and R.C. Williamson, "Hyperkernels," in *Advances in Neural Information Processing Systems*, Vancouver and Whistler, Canada, Dec. 8–13, vol. 15, 2003.
62. C. S. Ong, A. Smola, and B. Williamson, "Learning the kernel with hyperkernels," *J. Mach. Learn. Res.*, vol. 6, pp. 1045–1071, 2005.
63. L. Jia and S. Liao, "Hyperkernel construction for support vector machines," in *Proc. 4th Intl. Conf. Natural Computation*, Jinan, Shandong, China, Oct. 18–20, vol. 2, pp. 76–80, 2008.
64. J. Sun, et al., "Analysis of the distance between two classes for tuning SVM hyperparameters," *IEEE Trans. Neural Netw.*, vol. 21, no. 2, pp. 305–318, 2010.
65. C. Cortes, et al., "Algorithms for learning kernels based on centered alignment," *J. Mach. Learn. Res.*, vol. 13, pp. 795–828, 2012.
66. C. Burges, "A tutorial on support vector machines for pattern recognition," *Data Min. Knowl. Disc.*, vol. 2, no. 2, pp. 121–167, 1998.
67. B. Scholkopf and A. J. Smola, *Learning with kernels*. Cambridge, MA: MIT Press, 2002.
68. A. Argyriou, R. Hauser, C. A. Micchelli, and M. Pontil, "A DC-programming algorithm for kernel selection," in *Proc. 23rd Int. Conf. Machine Learning*, Pittsburgh, Pennsylvania, June 25–29, pp. 41–48, 2006.
69. S.-J. Kim, A. Magnani, and S. Boyd, "Optimal kernel selection in kernel Fisher discriminant analysis," in *Proc. 23rd Int. Conf. Machine Learning*, Pittsburgh, Pennsylvania, June 25–29, pp. 465–472, 2006.
70. R. Khemchandani and J. S. Chandra, "Optimal kernel selection in twin support vector machines," *Optim. Lett.*, vol. 3, no. 1, pp. 77–88, 2009.
71. R. Khemchandani and J. S. Chandra, "Learning the optimal kernel for fisher discriminant analysis via second order cone programming," *Eur. J. Oper. Res.*, vol. 203, no. 3, pp. 692–697, 2010.
72. S. R. Gunn and J. S. Kandola, "Structural modelling with sparse kernels," *Mach. Learn.*, vol. 48, pp. 137–163, 2002.
73. T. Jebara, "Multi-task feature and kernel selection for SVMs," in *Proc. 21st Int. Conf. Machine Learning*, Banff, Alberta, Canada, July 4–8, 2004.
74. K. Chaudhuri, C. Monteleoni, and A.D. Sarwate, "Differentially private empirical risk minimization," *J. Mach. Learn. Res.*, vol.12, pp. 1069–1109, 2011.
75. J. Neumann, C. Schnörr, G. Steidl, "SVM-based feature selection by direct objective minimisation," *Pattern Recogn. Lec. Notes Comput. Sci.*, vol. 3175, pp. 212–219, 2004.
76. Y. Lin and H. H. Zhang, "Component selection and smoothing in smoothing spline analysis of variance models," COSSO. Institute of Statistics MIMEO Series 2556, NCSU, January 2003.
77. C.A. Micchelli and M. Pontil, "Learning the kernel function via regularization," *J. Mach. Learn. Res.*, vol. 6, pp. 1099–1125, 2005.
78. A. Argyriou, C. A. Micchelli, and M. Pontil, "Learning convex combinations of continuously parameterized basic kernels," COLT'05 in *Proc. 18th Conf. Learning Theory*, Bertinoro, Italy, June 27–30, pp. 338–352, 2005.
79. F. R. Bach, and G. R. G. Lanckriet, "Multiple kernel learning, conic duality, and the SMO algorithm," in *Proc. 21th Int. Conf. Machine Learning*, Banff, Alberta, Canada, July 4–8, pp. 6–13, 2004.

80. S. Sonnenburg, G. Rätsch, and C. Schäfer, "A general and efficient multiple kernel learning algorithm," *Advances in Neural Information Processing Systems*, Vancouver and Whistler, Cananda, Dec. 4–7, vol. 18, pp. 1273–1280, 2006.
81. G. Dai and D. Yeung, "Kernel selection for semi-supervised kernel machines," in *Proc. 24th Int. Conf. Machine Learning*, Corvallis, OR, June 20–24, pp. 185–192, 2007.
82. S. O. Haykin, "Models of a neuron," *Neural networks and learning machines*. 3rd ed. New York: Macmillan College Publishing Company, pp. 8–13, 1994.
83. G. Camps-Valls and L. Bruzzone, "Kernel-based methods for hyperspectral image classification," *IEEE Trans. Geosci. Remote Sens.*, vol. 43, no. 6, pp. 1351–1362, 2005.
84. F. Melgani, "Classification of hyperspectral remote sensing images with support vector machines," *IEEE Trans. Geosci. Remote Sens.*, vol. 42, no. 8, pp. 1778–1790, 2004.
85. S. Donniger, T. Hofmann, and J. Yeh. "Predicting CNS permeability of drug molecules: Comparison of neural network and support vector machine algorithms," *J. Comput. Biol.*, vol. 9, no. 6, pp. 849–864, 2002.
86. D. West "Neural network credit scoring models," *Comput. Oper. Res.*, vol. 27, no. 11–12, pp. 1131–1152, 2000.
87. Y. J. Oyang, S. C. Hwang, Y. Y. Ou, C. Y. Chen, and Z. W. Chen, "Data classification with radial basis function networks based on a novel kernel density estimation algorithm," *IEEE Trans. Neural Netw.*, vol. 16, no. 1, pp. 225–236, 2005.
88. K. Reddy and V. Ravi, "Differential evolution trained kernel principal component WNN and kernel binary quantile regression: Application to banking," *Knowl. Based Syst.*, vol. 39, pp. 45–56, 2012.
89. Y. Xiao and H. Yigang. "A novel approach for analog fault diagnosis based on neural networks and improved kernel PCA," *Neurocomputing*, vol. 74, no. 7, pp. 1102–1115, 2011.
90. D. X. Niu, Q. Wang, and J. C. Li. "Short term load forecasting model using support vector machine based on artificial neural network," in *Proc. 4th Intl. Conf. Machine Learning and Cybernetics*, Guangzhou, China, August 18–21, vol. 1–9, pp. 4260–4265, 2005.
91. D. C. Park, N. H. Tran, D. M. Woo, and Y. Lee. "Kernel-based centroid neural network with spatial constraints for image segmentation," in *Proc. IEEE 4th Intl. Conf. Natural Computation*, Jinan, Shandong, China, Oct. 18–20, vol. 3, pp. 236–240, 2008.
92. Y. G. Xiao and L. G. Feng, "A novel neural-network approach of analog fault diagnosis based on kernel discriminant analysis and particle swarm optimization," *Appl. Soft Comput.*, vol. 12, no. 2, pp. 904–920, 2012.
93. A. Micheli, F. Portera, and A. Sperduti, "A preliminary empirical comparison of recursive neural networks and tree kernel methods on regression tasks for tree structured domains," *Neurocomputing*, vol. 64, pp. 73–92, 2005.
94. Y. Xia and J. Wang, "A one-layer recurrent neural network for support vector machine learning," *IEEE Trans. Syst. Man Cyb. B*, vol. 34, no. 2, pp. 1261–1269, 2004.
95. K. H. Jang, T. K. Yoo, J. Y. Choi, K. C. Nam, J. L. Choi, M. L. Kwon, and D. W. Kim, "Comparison of survival predictions for rats with hemorrhagic shocks using an artificial neural network and support vector machine," in *33rd Intl. Conf. EMBS*, Boston, Massachusetts, USA, August–September 2011.
96. C. Huang, L. S. David, and J. R. G. Townshend, "An assessment of support vector machines for land cover classification," *J. Remote Sens.*, vol. 23, no. 4, pp. 725–749, 2002.

97. K. A. Aziz, S. S. Abdullah, R. A. Ramlee, and A. N. Jahari, "Face detection using radial basis function neural networks with variance spread value," in *Intl. Conf. Soft Computing and Pattern Recognition*, Malacca, Malaysia, Dec. 4–7, 2009.
98. R. Moraes, J. F. Valiati, and W. P. G. Neto, "Document-level sentiment classification: An empirical comparison between SVM and ANN," *Expert Syst. Appl.*, vol. 40, pp. 621–633, 2013.
99. S. Arora, D. Bhattacharjee, M. Nasipuri, L. Malik, M. Kundu, and D. K. Basu, "Performance comparison of SVM and ANN for handwritten Devnagari character recognition," *J. Comput. Sci. Iss.* vol. 7, no. 3, 2010.
100. R. Burbidge, M. Trotter, B. Buxton, and S. Holden, "Drug design by machine learning support vector machines for pharmaceutical analysis," *Comput. Chem.*, vol. 26, no. 1, pp. 5–14, 2001.
101. A. Statnikov, C. F. Aliferis, I. Tsamardions, D. Hardin, and S. Levy, "A comprehensive evaluation of multicategory classification methods for microarray gene expression cancer diagnosis," *Bioinformatics*, vol. 21 no. 5, pp. 631–643, 2005.
102. V. N. Vapnik, "An overview of statistical learning theory," *IEEE Trans. Neural Netw.*, vol. 10, no 5, pp. 988–999, 1999.
103. B. Boser I. Guyon, and V. Vapnik, "A training algorithm for optimal margin classifier," in *Proc. 5th ACM Workshop Computational Learning Theory*, Pittsburgh, PA, July 27–29, pp. 144–152, 1992.
104. M. Pontil and A. Verri, "Properties of support vector machines," *Neural Comput.*, vol. 10, no. 4, pp. 955–974, 1998.
105. C. Campbell and Y. Ying, *Learning with support vector machines*. San Rafael, CA: Morgan & Claypool, pp. 6, 2011.
106. V.D. Sanchez A, "Advanced support vector machines and kernel methods," *Neurocomputing*, vol. 55, pp. 5–20, 2003.
107. J. Shawe-Taylor and S. Sun, "A review of optimization methodologies in support vector machines," *Neurocomputing*, vol. 74, no. 17, pp. 3609–3618, 2011.
108. J. C. Platt, "Fast training of support vector machines using sequential minimal optimization," in *Advances in kernel methods—Support vector learning*, B. Scholkopf, C.J.C. Burges, and A.J Smola (Eds.). Cambridge, MA: The MIT Press, 1998.
109. G. W. Flake and S. Lawrence, "Efficient SVM regression training with SMO," *Mach. Learn.*, vol.46, pp. 271–290, 2002.
110. P. Laskov, "An improved decomposition algorithm for regression support vector machines," *Mach. Learn.*, 46, pp. 315–350, 2002.
111. E. Osuna, R. Freund, and F. Girosi, "An improved training algorithm for support vector machines," in *Proc. of IEEE Neural Networks and Signal Processing*, Amelia Island, FL, Sep. 24–26, 1997.
112. O. L. Mangasarian and D. R. Musicant, "Successive overrelaxation for support vector machines," *IEEE Trans. Neural Netw.*, vol.10, no.5, pp. 1032–1037, 1999.
113. T. Friess, N. Cristianini, and C. Campbell, "The Kernel-adatron: A fast and simple learning procedure for support vector machines," in *Proc. 5th Intl. Conf. Machine Learning*, Helsinki, Finland, June 5–9, pp. 188–196, 1998.
114. A. J. Smola and B. Scholkopf, "From regularization operators to support vector kernels," in *Advances in Neural Information Processing Systems*, Denver, CO, Dec. 1–3, vol. 10, pp. 343–349, 1998.

115. D. Meyer, F. Leisch, and K. Hornik, "The support vector machine under test," *Neurocomputing*, vol. 55, no. 1–2, pp. 169–186, 2003.

116. K. Jayadeva, R. Khemchandani, and S. Chandra, "Twin support vector machines for pattern classification," *IEEE Trans. Pattern Anal. Mach. Intell.*, vol.29, no.5, pp. 905–910, 2007.

117. M. Arun Kumar and M. Gopal, "Least squares twin support vector machines for pattern classification," *Expert Syst. Appl.*, vol.36, no. 4, pp. 7535–7543, 2009.

118. B. Schölkopf, A. Smola, and K.-R. Müller, "Nonlinear component analysis as a kernel eigenvalue problem," *Neural Comput.*, vol. 10, no. 5, pp. 1299–1319, 1998.

119. E. Barshan, A. Ghodsi, and Z. Azimifar, "Supervised principal component analysis: Visualization, classification and regression on subspaces and submanifolds," *Pattern Recogn.*, vol. 44, no. 7, pp. 1357–1371, 2011.

120. J. Yang, A. F. Frangi, and J. Y. Yang, "KPCA plus LDA: A complete kernel fisher discriminant framework for feature extraction and recognition," *IEEE Trans. Pattern Anal. Mach. Intell.*, vol. 27, no. 2, pp. 230–244, 2005.

121. C. J. Liu, "Gabor-based kernel PCA with fractional power polynomial models for face recognition," *IEEE Trans. Pattern Anal. Mach. Intell.*, vol. 26, no. 5, pp. 572–581, 2004.

122. S. Lemm, B. Blankertz, T. Dickhaus, and K.-R. Müller "Introduction to machine learning for brain imaging," *Neuroimage*, vol. 56, no. 2, pp. 387–399, May 2011.

123. R. M. Balabin and R. Z. Safieva, "Near-infrared (NIR) spectroscopy for biodiesel analysis: Fractional composition, iodine value, and cold filter plugging point from one vibrational spectrum," *Energy Fuel* vol. 25, no. 5, pp. 2373–2382, 2011.

124. X. Yang, S. Gao, S. Xu, Z. Zhang, B. M. Prasanna, L. Li, J. Li, and J. Yan, "Characterization of a global germplasm collection and its potential utilization for analysis of complex quantitative traits in maize," *Mol. Breeding*, vol. 28, no. 4, pp 511–526, 2011.

125. J. Yu, "Nonlinear bioprocess monitoring using multiway kernel localized fisher discriminant analysis," *Indus. Eng. Chem. Res.*, vol. 50, no. 6, pp. 3390–3402, 2011.

126. Y. Zhang and C. Ma, "Fault diagnosis of nonlinear processes using multiscale KPCA and multiscale KPLS," *Chem. Eng. Sci.*, vol. 66, no. 1, pp. 64–72, 2011.

127. J. T. Leek, "Asymptotic conditional singular value decomposition for high-dimensional genomic data," *Biometrics*, vol. 67, no. 2, pp. 344–352, 2011.

128. G. Liu, X. Sun, and K. Fu, "Aircraft recognition in high-resolution satellite images using coarse-to-fine shape prior," *IEEE Geosci. Remote Sens. Lett.*, vol. 10, no. 3, pp. 573–577, May 2013.

129. A. Erener, "Classification method, spectral diversity, band combination and accuracy assessment evaluation for urban feature detection," *J. Appl. Earth Obs. Geoinform.*, vol. 21, pp. 397–408, 2013.

130. H. Liu, M. Huang, and J. T. Kim, "Adaptive neuro-fuzzy inference system based faulty sensor monitoring of indoor air quality in a subway station," *Kor. J. Chem. Eng.*, vol. 30, no. 3, pp. 528–539, 2013.

131. W.-Y. Lin and C.-Y. Hsieh, "Kernel-based representation for 2D/3D motion trajectory retrieval and classification," *Pattern Recogn.*, vol. 46, no. 3, pp. 662–670, 2013.

132. G. Rong, S.-Y. Liu, and J.-D. Shao, "Fault diagnosis by locality preserving discriminant analysis and its kernel variation," *Chem. Eng. Sci.*, vol. 49, pp. 105–113, 2013.

133. K. N. Reddy, and V. Ravi, "Differential evolution trained kernel principal component WNN and kernel binary quantile regression: Application to banking," *Knowl. Based Syst.*, vol. 39, pp. 45–56, 2013.

134. S. Liwicki, G. Tzimiropoulos, and S. Zafeiriou, "Euler principal component analysis," *J. Comput. Vision*, vol. 101, no. 3, pp. 498–518, 2013.

135. D. K. Saxena, J. A. Duro, and A. Tiwari, "Objective reduction in many-objective optimization: Linear and nonlinear algorithms," *IEEE Trans. Evol. Comput.*, vol. 17, no. 1, pp. 77–99, 2013.

136. C. Silva, U. Lotric, B. Ribeiro, A. Dobnikar, "Distributed text classification with an ensemble kernel-based learning approach," *IEEE Trans. Syst. Man Cyb. C*, vol. 40, no.3, pp. 287-297, May 2010.

137. T. Gärtner, "A survey of kernels for structured data," *SIGKDD Explor. Newsl.*, vol. 5, no. 1, pp 49–58, 2003.

138. R. Rosipal and L. J. Trejo, "Kernel partial least squares regression in reproducing kernel Hilbert space," *J. Mach. Learn. Res.*, vol. 2, pp. 97–123, 2002

139. D. Hardoon, S. Szedmak, and J. Shawe-Taylor, "Canonical correlation analysis: An overview with application to learning methods," *Neural Comput.*, vol. 16, no. 12, pp. 2639–2664, 2004.

140. R.I. Kondor and J. Lafferty, "Diffusion kernels on graphs and other discrete input spaces," in *Proceeding ICML '02 Proceedings of the Nineteenth International Conference on Machine Learning*, Sydney, NSW, Australia, July 8–12, vol. 2, pp. 315–322, 2002.

141. B. Yifeng, X. Jian, and Y. Long. "Kernel partial least-squares regression," in *Neural Networks, IJCNN'06. International Joint Conference on IEEE*, Vancouver, BC, Canada, July 16–21, 2006.

142. S. Wold, et al., "The collinearity problem in linear regression. The partial least squares (PLS) approach to generalized inverses," *SIAM J. Sci. Statist. Comput.*, vol.5, no.3, pp. 735–743, 1984.

143. S. de Jong, "SIMPLS: An alternative approach to partial least squares regression," *Chemometr. Intell. Lab. Syst.*, vol. 18, no. 3, pp. 251–263, 1993.

144. K. Kim, J. M. Lee, and I. B. Lee, "A novel multivariate regression approach based on kernel partial least squares with orthogonal signal correction," *Chemometr. Intell. Lab. Syst.*, vol. 79, no. 1, pp. 22–30, 2005.

145. W. H. A. M.V.D. Broek, et al., "Plastic identification by remote sensing spectroscopic NIR imaging using kernel partial least squares (KPLS)," *Chemometr. Intell. Lab. Syst.*, vol. 35, no. 2, pp 187–197, 1996.

146. V. Štruc, and N. Pavešić, "Gabor-based kernel partial-least-squares discrimination features for face recognition," *Informatica*, vol. 20, no.1, pp. 115–138, 2009.

147. Lai, P. L., and C. Fyfe, "Kernel and nonlinear canonical correlation analysis," *Int. J. Neural Syst.*, vol. 10, no. 5, pp. 365, 2000.

148. T. Melzer, M. Reiter, and H. Bischof, "Appearance models based on kernel canonical correlation analysis," *Pattern Recogn.*, vol. 36, no.9, pp. 1961–1971, 2003.

149. W. Zheng, et al., "Facial expression recognition using kernel canonical correlation analysis (KCCA)," *IEEE Trans. Neural Netw.*, vol.17, no.1, pp 233–238, 2006.

150. Y. Yamanishi, et al., "Extraction of correlated gene clusters from multiple genomic data by generalized kernel canonical correlation analysis," *Bioinformatics*, vol. 19, no. 1, pp. 323–330, 2003.

151. K. Fukumizu, F. R. Bach, and A. Gretton. "Statistical consistency of kernel canonical correlation analysis," *J. Mach. Learn. Res.*, vol. 8, pp. 361–383, 2007.

152. E. Kokiopoulou, and Y. Saad. "Orthogonal neighborhood preserving projections: A projection-based dimensionality reduction technique," *IEEE Trans. Pattern Anal. Mach. Intell.*, vol. 29, no.12, pp 2143–2156, 2007.

153. D.R. Hardoon and J. Shawe-Taylor, "Convergence analysis of kernel canonical correlation analysis: Theory and practice," *Mach. Learn.*, vol. 74, no. 1, pp. 23–38, 2009.

154. G. Boente, and R. Fraiman. "Kernel-based functional principal components," *Stat. Prob. Let.*, vol. 48, no.4, pp. 335–345, 2000.

155. D. Cai, et al., "Orthogonal laplacianfaces for face recognition," *IEEE Trans. Image Processing*, vol. 15, no.11, pp. 3608–3614, 2006.

156. A. Ruiz and P. E. López-de-Teruel, "Nonlinear kernel-based statistical pattern analysis," *IEEE Trans. Neural Netw.*, vol. 12, no.1, pp. 16–32, 2001.

157. J. Yang, et al. "Globally maximizing, locally minimizing: Unsupervised discriminant projection with applications to face and palm biometrics," *IEEE Trans. Pattern Anal. Mach. Intell.*, vol. 29, no.4, pp. 650–664, 2007.

158. J. Arenas-Garcia, K. Petersen, G. Camps-Valls, and L.K. Hansen, "Kernel multivariate analysis framework for supervised subspace learning: A tutorial on linear and kernel multivariate methods," *IEEE Signal Process. Mag.*, vol. 30, no. 4, pp. 16,29, 2013

159. J. Aitchison and C. GG Aitken, "Multivariate binary discrimination by the kernel method," *Biometrika*, vol.63, no.3, pp. 413–420, 1976.

160. E. Pereda, R. Q. Quiroga, and J. Bhattacharya, "Nonlinear multivariate analysis of neurophysiological signals," *Prog. Neurobiol.*, vol. 77, no.1, pp. 1–37, 2005.

161. B.A. Weinstock, et al. "Prediction of oil and oleic acid concentrations in individual corn (*Zea mays* L.) kernels using near-infrared reflectance hyperspectral imaging and multivariate analysis," *Appl. Spectrosc.*, vol. 60, no.1, pp. 9–16, 2006.

162. Z. D. Bai, C. R. Rao, and L. C. Zhao. "Kernel estimators of density function of directional data," *J. Multivariate Anal.*, vol. 27, no.1, pp. 24–39, 1988.

163. T. Duong, "k_s: Kernel density estimation and kernel discriminant analysis for multivariate data in R," *J. Stat. Software*, vol. 21, no.7, pp. 1–16, 2007.

164. D. K. Hammond, P. Vandergheynst, and R. Gribonval. "Wavelets on graphs via spectral graph theory," *Appl. Comput. Harmonic Anal.*, vol. 30, no. 2, pp. 129–150, 2011.

165. S. Köhler, et al. "Walking the interactome for prioritization of candidate disease genes," *Am. J. Human Genet.*, vol. 82, no.4, pp. 949–958, 2008.

166. M. Bylesjö, et al., "K-OPLS package: Kernel-based orthogonal projections to latent structures for prediction and interpretation in feature space," *BMC Bioinformatics*, vol. 9, no. 1, pp. 106, 2008.

167. M. Rantalainen, et al., "Kernel-based orthogonal projections to latent structures (K-OPLS)," *J. Chemometr.*, vol. 21, no. 7–9, pp. 376–385, 2007.

168. S. Zhao, F. Precioso, M. Cord, "Spatio-temporal tube data representation and kernel design for SVM-based video object retrieval system," *Multimed. Tools Appl.*, vol. 55, no. 1, pp. 105–125, 2011.

169. F. Dinuzzo, G. Pillonetto, and G.D. Nicolao, "Client–server multitask learning from distributed datasets," *IEEE Trans. Neural Netw.*, vol. 22 no. 2, 2011.

170. R. Caruana, "Multitask learning," *Mach. Learn.*, vol. 28 issue 1, pp. 41–75, 1997.

171. A. Kitamoto, "Typhoon analysis and data mining with kernel methods," *Pattern recognition with support vector machines*. Berlin, Heidelberg: Springer, pp. 237–249, 2002.
172. B. L. Milenova, J. S. Yarmus, M. M. Campos, "SVM in oracle database 10g: Removing the barriers to widespread adoption of support vector machines," in *Proc. 31st Int. Conf. Very Large Databases*, VLDB Endowment, Trondheim, Norway, August 30–September 2, 2005.
173. Y. Zhang, "Enhanced statistical analysis of nonlinear processes using KPCA, KICA and SVM," *Chem. Eng. Sci.*, vol. 64, no. 5, pp. 801–811, 2009.
174. J. X. Dong, A. Krzyzak, and C. Y. Suen, "Fast SVM training algorithm with decomposition on very large data sets," *IEEE Trans. Pattern Anal. Mach. Intell.*, vol. 27, no. 4, pp. 603–618, 2005.
175. C. J. Lin, "Large-scale machine learning in distributed environments." Tutorial at *ACM ICMR*. Jun 2012. [Online]. Available: http://www.csie.ntu.edu.tw/~cjlin/talks/icmr2012.pdf. Accessed 18 November 2014.
176. Z.A. Zhu, W. Z. Chen, G. Wang, C. G. Zhu, and Z. Chen, "P-packSVM: Parallel primal gradient descent Kernel SVM," in *9th IEEE Intl Conf. Data Mining*. [Online]. http://people.csail.mit.edu/zeyuan/paper/2009-ICDM-Parallel.pdf. Accessed 18 November 2014.
177. E.Y. Chang, K. Zhu, H. Wang, and H. Bai, "PSVM: Parallelizing support vector machines on distributed computers," *Adv. Neural Inform. Process. Syst.*, vol. 20, pp. 257–264, 2008.
178. Z. Y. Chen, Z. P. Fan, and M. Sun, "A hierarchical multiple kernel support vector machine for customer churn prediction using longitudinal behavioral data," *Eur. J. Oper. Res.*, vol. 223, pp. 461–472, 2012.
179. N. A. Syed, H. Liu, and K. K. Sung, "Handling concept drifts in incremental learning with support vector machines," in *5th ACM SIGKDD Intl. Conf. Knowledge Discovery and Data Mining*, San Diego, CA, USA August 15–18, 1999.
180. Z. Liang and Y. Li, "Incremental support vector machine learning in the primal and applications." *Neurocomputing*, vol. 72, pp. 2249–2258, 2009.
181. S. Fine and K. Scheinberg, "Efficient SVM training using low-rank kernel representations." *J. Mach. Learn. Res.*, vol. 2, pp. 243–264, 2001.
182. H. Xu, Y. Wen, and J. Wang, "A fast-convergence distributed support vector machine in small-scale strongly connected networks." *Front. Electr. Electron. Eng.*, vol. 7, no. 2, pp. 216–223, 2011.
183. T. T. Frieb, N. Cristianini, and C. Campbell, "The kernel-adatron algorithm: A fast and simple learning procedure for support vector machines," in *Proc. 15th Intl. Conf. Machine Learning*, 1998.
184. K. Woodsend, G. Jacek, "Exploiting separability in large-scale linear support vector machine training," *Comput. Optim. Appl.*, vol. 49, no. 2, pp. 241–269, 2011.
185. Z. Chen, Z. Fan, "Dynamic customer lifetime value prediction using longitudinal data: An improved multiple kernel SVR approach," *Knowl. Based Syst.*, vol. 43, pp. 123–134, 2013.
186. Z.-Y. Chen and Z.-P. Fan, "Distributed customer behavior prediction using multiplex data: A collaborative MK-SVM approach," *Knowl. Based Syst.*, vol.35, pp. 111–119, 2012.
187. Z. Chen, J. Li, L. Wei, W. Xu, and Y. Shi, "Multiple-kernel SVM based multiple-task oriented data mining system for gene expression data analysis," *Expert Syst. Appl.*, vol. 38, no. 10, pp. 12151–12159, 2011.

188. J. Cheng, J. Qian, Y. Guo, "A distributed support vector machines architecture for chaotic time series prediction," *Neural Inform. Process.*, vol. 4232, pp 892–899, 2006.

189. C. Blake, E. Keogh, C. Merz, *UCI repository of machine learning databases.* [Online] http://archive.ics.uci.edu/ml/, 1998.

190. T. Rose, M. Stevenson, and M. Whitehead, "The reuters corpus volume 1-from yesterday's news to tomorrow's language resources," *LREC*, Vol. 2, pp. 827–832, 2002.

191. S. Agarwal, V. Vijaya Saradhi, and H. Karnick, "Kernel-based online machine learning and support vector reduction," *Neurocomputing*, vol. 71, no. 7–9, pp. 1230–1237, 2008.

192. A. Singh, N. Ahuja, and P. Moulin, "Online learning with kernels: Overcoming the growing sum problem," in *Proc. IEEE Intl Workshop on Machine Learning for Signal Processing*, Santander, Spain, Sep. 23–26, pp. 1–6, 2012.

193. F. Orabona, J. Keshet, and B. Caputo, "Bounded kernel-based online learning," *J. Mach. Learn. Res.*, vol. 10, pp. 2643–2666, Dec. 2009.

194. K. Slavakis, S. Theodoridis, and I. Yamada, "Adaptive constrained learning in reproducing kernel hilbert spaces: The robust beamforming case," *IEEE Trans. Signal Processing*, vol. 57, no. 12, pp. 4744–4764, 2009.

195. Y. Engel, S. Mannor, and R. Meir, "The kernel recursive least-squares algorithm," *IEEE Trans. Signal Processing*, vol. 52, no. 8, pp. 2275–2285, 2004.

196. L. Csató and M. Opper, "Sparse on-line Gaussian processes," *Neural Comput.*, vol. 14, no. 3, pp. 641–668, Mar. 2002.

197. B. Chen, S. Zhao, P. Zhu, and J. C. Principe, "Quantized kernel least mean square algorithm," *IEEE Trans. Neural Netw. Learn. Syst.*, vol. 23, no. 1, pp. 22–32, 2012.

198. R. Jin, S. C. H. Hoi, and T. Yang, "Online multiple kernel learning: algorithms and mistake bounds," in *Algorithmic learning theory*, M. Hutter, F. Stephan, V. Vovk, and T. Zeugmann, Eds. Berlin, Heidelberg: Springer, pp. 390–404, 2010.

199. S. C. H. Hoi, R. Jin, P. Zhao, and T. Yang, "Online multiple kernel classification," *Mach. Learn.*, vol. 90, no. 2, pp. 289–316, 2012.

200. F. Rosenblatt, "The perceptron: A probabilistic model for information storage and organization in the brain," *Psychol. Rev.*, vol. 65, no. 6, pp. 386–408, 1958.

201. J. P. Rhinelander and X. X. Lu, "Stochastic subset selection for learning with kernel machines," *IEEE Trans. Syst. Man Cyb. B*, vol. 42, no. 3, pp. 616–626, 2012.

202. I. Basheer and M. Hajmeer, "Artificial neural networks: Fundamentals, computing, design, and application," *J. Microbiol. Meth.*, vol. 43, no. 1, pp. 3–31, 2000.

203. L. Rutkowski, "Adaptive probabilistic neural networks for pattern classification in time-varying environment," *IEEE Trans. Neural Netw.*, vol. 15, no. 4, pp. 811–827, 2004.

204. Y.-H. Liu, Y.-C. Liu, and Y.-J. Chen, "Fast support vector data descriptions for novelty detection," *IEEE Trans. Neural Netw.*, vol. 21, no. 8, pp. 1296–1313, 2010.

205. H. Hoffmann, "Kernel PCA for novelty detection," *Pattern Recogn.*, vol. 40, no. 3, pp. 863–874, 2007.

206. A. M. Jade, B. Srikanth, V. K. Jayaraman, B. D. Kulkarni, J. P. Jog, and L. Priya, "Feature extraction and denoising using kernel PCA," *Chem. Eng. Sci.*, vol. 58, no. 19, pp. 4441–4448, 2003.

207. K. I. Kim, M. O. Franz, and B. Scholkopf, "Iterative kernel principal component analysis for image modeling," *IEEE Trans. Pattern Anal. Mach. Intell.*, vol. 27, no. 9, pp. 1351–1366, 2005.

208. C. Gruber, T. Gruber, S. Krinninger, and B. Sick, "Online signature verification with support vector machines based on LCSS kernel functions," *IEEE Trans. Syst. Man Cyb. B*, vol. 40, no. 4, pp. 1088–1100, 2010.
209. M. Davy, F. Desobry, A. Gretton, and C. Doncarli, "An online support vector machine for abnormal events detection," *Signal Processing*, vol. 86, no. 8, pp. 2009–2025, 2006.
210. F. Desobry, M. Davy, and C. Doncarli, "An online kernel change detection algorithm," *IEEE Trans. Signal Processing*, vol. 53, no. 8, pp. 2961–2974, 2005.
211. Y. Zhang, N. Meratnia, and P. Havinga, "Adaptive and online one-class support vector machine-based outlier detection techniques for wireless sensor networks," in *Proc. Intl. Conf. Advanced Information Networking and Applications Workshops*, Bradford, United Kingdom, May 26–29, pp. 990–995, 2009.
212. V. Gomez-Verdejo, J. Arenas-Garcia, M. Lazaro-Gredilla, and A. Navia-Vazquez, "Adaptive one-class support vector machine," *IEEE Trans. Signal Processing*, vol. 59, no. 6, pp. 2975–2981, 2011.
213. K. Crammer, O. Dekel, J. Keshet, S. Shalev-Shwartz, and Y. Singer, "Online passive-aggressive algorithms," *J. Mach. Learn. Res.*, vol. 7, pp. 551–585, 2006.
214. C. Richard, J. C. M. Bermudez, and P. Honeine, "Online prediction of time series data with kernels," *IEEE Trans. Signal Processing*, vol. 57, no. 3, pp. 1058–1067, 2009.
215. W. Wang, C. Men, and W. Lu, "Online prediction model based on support vector machine," *Neurocomputing*, vol. 71, no. 4–6, pp. 550–558, 2008.
216. K. Kuusilinna, V. Lahtinen, T. Hamalainen, and J. Saarinen, "Finite state machine encoding for VHDL synthesis," *IEE Proc. Comput. Digital Tech.*, vol. 148, no. 1, pp. 23–30, 2001.
217. M. Martinez-Ramon, J. L. Rojo-Alvarez, G. Camps-Valls, J. Munoz-Mari, A. Navia-Vazquez, E. Soria-Olivas, and A. R. Figueiras-Vidal, "Support vector machines for nonlinear kernel ARMA system identification," *IEEE Trans. Neural Netw.*, vol. 17, no. 6, pp. 1617–1622, 2006.
218. R. K. K. Yuen, E. W. M. Lee, C. P. Lim, and G. W. Y. Cheng, "Fusion of GRNN and FA for online noisy data regression," *Neural Process. Lett.*, vol. 19, no. 3, pp. 227–241, 2004.
219. H. A. Kingravi, G. Chowdhary, P. A. Vela, and E. N. Johnson, "A reproducing kernel Hilbert space approach for the online, update of radial bases in neuro-adaptive control," in *Proc. 50th IEEE Conf. Decision and Control and European Control*, Orlando, Florida, December 12–15, pp. 1796–1802, 2011.
220. J. Si and Y.-T. Wang, "Online learning control by association and reinforcement," *IEEE Trans. Neural Netw.*, vol. 12, no. 2, pp. 264 –276, 2001.
221. S. Marsland, "Novelty detection in learning systems," *Neural Comput. Surv.*, vol. 3, no. 2, pp. 157–195, 2003.
222. B. Sofman, B. Neuman, A. Stentz, and J. A. Bagnell, "Anytime online novelty and change detection for mobile robots," *J. Field Robotics*, vol. 28, no. 4, pp. 589–618, 2011.
223. O. Taouali, I. Elaissi, and H. Messaoud, "Online identification of nonlinear system using reduced kernel principal component analysis," *Neural Comput. Appl.*, vol. 21, no. 1, pp. 161–169, Feb. 2012.
224. S. Ozawa, Y. Takeuchi, and S. Abe, "A fast incremental kernel principal component analysis for online feature extraction," in *Proc. PRICAI 2010: Trends in Artificial Intelligence*, B.-T. Zhang and M. A. Orgun, Eds. Berlin Heidelberg: Springer, 2010, Daegu, Korea, August 30–September 2, pp. 487–497, 2010.

225. S. Salti and L. Di Stefano, "On-line support vector regression of the transition model for the Kalman filter," *Image Vision Comput.*, vol. 31, no. 6–7, pp. 487–501, 2013.
226. J. Makhoul, "Linear prediction: A tutorial review," *Proc. IEEE*, vol. 63, no. 4, pp. 561–580, 1975.
227. D. Wingate and S. Singh, "Kernel predictive linear Gaussian models for nonlinear stochastic dynamical systems," in *Proc. Intl. Conf. Machine Learning*, Pittsburgh, Pennsylvania, June 25–29, pp. 1017–1024, 2006.
228. G. Welch and G. Bishop, "An introduction to the Kalman filter," 1995. [Online]. http://clubs.ens-cachan.fr/krobot/old/data/positionnement/kalman.pdf. Accessed 17 Nov 2014.
229. R. S. Liptser and A. N. Shiryaev, *Statistics of random processes: II. Applications*. New York, NY: Springer, 2001.
230. E. D. Sontag, *Mathematical control theory: Deterministic finite dimensional systems*. New York, NY: Springer, 1998.
231. J. Durbin and S. Koopman, *Time series analysis by state space methods*, 2nd ed. Oxford, United Kingdom: Oxford University Press, 2012.
232. C. K. Chui and G. Chen, *Kalman filtering: With real-time applications*. New York, NY: Springer, 2009.
233. L. Ralaivola and F. d' Alche-Buc, "Time series filtering, smoothing and learning using the kernel Kalman filter," in *Proc. Intl. Joint Conf. Neural Networks*, Montreal, QC, Canada, July 31–August 4, vols. 1–5, pp. 1449–1454, 2005.
234. F. M. Waltz and J. W. V. Miller, "Gray-scale image processing algorithms using finite-state machine concepts," *J. Electron. Imaging*, vol. 10, no. 1, pp. 297–307, 2001.
235. C. Saunders, J. Shawe-Taylor, and A. Vinokourov, "String kernels, Fisher kernels and finite state automata," in *Advances in Neural Information Processing Systems*, Vancouver, Whistler, Canada, Dec. 8–13, vol. 15, pp. 633–640, 2003.
236. T. Natschläger and W. Maass, "Spiking neurons and the induction of finite state machines," *Theor. Comput. Sci.*, vol. 287, no. 1, pp. 251–265, 2002.
237. L. Shpigelman, H. Lalazar, and E. Vaadia, "Kernel-ARMA for hand tracking and brain-machine interfacing during 3D motor control," in *Proc. NIPS*, Vancouver, B.C., Canada, December 8–11, pp. 1489–1496, 2008.
238. J. G. de Gooijer, B. Abraham, A. Gould, and L. Robinson, "Methods for determining the order of an autoregressive-moving average process: A survey," *Int. Stat. Rev./Revue Internationale de Statistique*, vol. 53, no. 3, pp. 301–329, Dec. 1985.
239. L. Hoegaertsa, L. De Lathauwerb, I. Goethalsa, J.A.K. Suykensa, J. Vandewallea, and B. De Moora, "Efficiently updating and tracking the dominant kernel principal components," *Neural Netw.*, vol.20, no. 2, pp. 220-229, 2007.
240. S. R. Upadhyaya, "Parallel approaches to machine learning—A comprehensive survey," *J. Parallel Distr. Com.*, vol.73, no. 3, pp. 284–292, 2013.
241. P. P. Pokharel, W. Liu, and J. C. Principe, "Kernel least mean square algorithm with constrained growth," *Signal Processing*, vol. 89, no. 3, pp. 257–265, 2009.
242. Y. Qi, P. Wang, S. Fan, X. Gao, and J. Jiang, "Enhanced batch process monitoring using Kalman filter and multiway kernel principal component analysis," in *Proc. 21st Control and Decision Conference*, Guilin, China, Jun. 17–19, vol. 5, pp. 5289–5294, 2009.

2

OFFLINE KERNEL ANALYSIS[1]

2.1 INTRODUCTION

Capitalizing on the recent success of kernel methods in pattern classification [1–4], Schölkopf and Smola [5] developed and studied a feature selection algorithm, in which principal component analysis (PCA) was effectively applied to a sample of n, d-dimensional patterns that are first injected into a high-dimensional Hilbert space using a nonlinear embedding. Heuristically, embedding input patterns into a high-dimensional space may elucidate salient, nonlinear features in the input distribution, in the same way that nonlinearly separable classification problems may become linearly separable in higher dimensional spaces as suggested by the Vapnik–Chervonenkis theory [6]. Both the PCA and the nonlinear embedding are facilitated by a Mercer kernel of two arguments $k : R^d \times R^d \rightarrow R$, which effectively computes the inner product of the transformed arguments. This algorithm, called kernel principal component analysis (KPCA), thus avoids the problem of representing transformed vectors in the Hilbert space and enables the computation of the inner product of two transformed vectors of an arbitrarily high dimension in constant time. Nevertheless, KPCA has two deficiencies: (i) the computation of the principal components involves the solution of an eigenvalue problem that requires $O(n^3)$ computations, and (ii) each principal component in the Hilbert space depends on every one of the n input patterns, which defeats the goal of obtaining both an informative and concise representation.

Both of these deficiencies have been addressed in subsequent investigations that seek sets of salient features that depend only on sparse subsets of transformed input patterns. Tipping [7] applied a maximum-likelihood technique to approximate the

[1]This chapter is a revised version of the author's paper in IEEE Transactions on Knowledge and Data Engineering, DOI: 10.1109/TKDE.2012.110, approved by IEEE Intellectual Property Rights.

Data-Variant Kernel Analysis, First Edition. Yuichi Motai.

transformed covariance matrix in terms of such a sparse subset. France and Hlaváč [8] proposed a greedy method, which approximates the mapped space representation by selecting a subset of input data. It iteratively extracts the data in the mapped space until the reconstruction error in the mapped, high-dimensional space falls below a threshold value. Its computational complexity is $O(nm^3)$, where n is the number of input patterns and m is the cardinality of the subset. Zheng [9] split the input data into M groups of similar size and then applied KPCA to each group. A set of eigenvectors was obtained for each group. KPCA was then applied to a subset of these eigenvectors to obtain a final set of features. Although these studies proposed useful approaches, none provided a method that is both computationally efficient and accurate.

To avoid the $O(n^3)$ eigenvalue problem, Mangasarian et al. [10] proposed sparse kernel feature analysis (SKFA), which extracts l features, one by one, using a l_1—constraint on the expansion coefficients. SKFA requires only $O(l^2n^2)$ operations and is thus a significant improvement over KPCA if the number of dominant features is much lesser than the data size. However, if $l > \sqrt{n}$, then the computational cost of SKFA is likely to exceed that of KPCA.

In this study, we propose an accelerated kernel feature analysis (AKFA) that generates l sparse features from a dataset of n patterns using $O(ln^2)$ operations. As AKFA is based on both KPCA and SKFA, we analyze the former algorithms, that is, KPCA and SKFA, and then describe AKFA in Section 2.2.

We have evaluated other existing multiple kernel learning approaches [11–13] and found that those approaches do not rely on the datasets to combine and choose the kernel functions very much. The choice of an appropriate kernel function has reflected prior knowledge concerning the problem at hand. However, it is often difficult for us to exploit the prior knowledge on patterns for choosing a kernel function, and how to choose the best kernel function for a given dataset is an open question. According to the no free lunch theorem [14] on machine learning, there is no superior kernel function in general, and the performance of a kernel function depends on applications, specifically the datasets. The five kernel functions, linear, polynomial, Gaussian, Laplace, and Sigmoid, are chosen because they were known to have good performances [7, 14–18].

The main contribution of this work is a principal composite kernel feature analysis (PC-KFA) described in Section 2.3. In this new approach, the kernel adaptation is employed in the kernel algorithms above KPCA and AKFA in the form of the best kernel selection, engineer a composite kernel that is a combination of data-dependent kernels, and the optimal number of kernel combination. Other multiple kernel learning approaches combined basic kernels, but our proposed PC-KFA specifically chooses data-dependent kernels as linear composites.

In Section 2.4, we summarize numerical evaluation experiments on the basis of medical image datasets (MIDs) in computer-aided diagnosis (CAD) using the proposed PC-KFA (i) to choose the kernel function, (ii) to evaluate feature representation by calculating reconstruction errors, (iii) to choose the number of kernel functions, (iv) to composite the multiple kernel functions, (v) to evaluate feature classification using a simple classifier, and (vi) to analyze the computation time. Our conclusions appear in Section 2.5.

2.2 KERNEL FEATURE ANALYSIS

2.2.1 Kernel Basics

Using Mercer's theorem [19], a nonlinear, positive-definite kernel $k : R^d \times R^d \rightarrow R$ of an integral operator can be computed by the inner product of the transformed vectors $\langle \Phi(x), \Phi(y) \rangle$, where $\Phi : R^d \rightarrow H$ denotes a nonlinear embedding (induced by k) into a possibly infinite dimensional Hilbert space H. Given n sample points in the domain $X_n = \{x_i \in R^d \mid i = 1, \ldots, n\}$, the image $Y_n = \{\Phi(x_i) \mid i = 1, \ldots, n\}$ of X_n spans a linear subspace of at most $(n-1)$ dimensions. By mapping the sample points into a higher dimensional space, H, the dominant linear correlations in the distribution of the image Y_n may elucidate important nonlinear dependencies in the original data sample X_n. This is beneficial because it permits making PCA nonlinear without complicating the original PCA algorithm. Let us introduce kernel matrix K as a Hermitian and positive, semidefinite matrix that computes the inner product between any finite sequences of inputs $x := \{x_j : j \in N_n\}$ and is defined as

$$K := (K(x_i, x_j) : i, j \in \mathrm{N}_n) = (\Phi(x_i) \cdot \Phi(x_j))$$

Commonly used kernel matrices are as follows [5]:

The linear kernel:

$$K(x, x_i) = x^T x_i \tag{2.1}$$

The polynomial kernel:

$$K(x, x_i) = (x^T x_i + \mathrm{Offset})^d \tag{2.2}$$

The Gaussian radial basis function (RBF) kernel:

$$K(x, x_i) = \exp\left(\frac{-\|x - x_i\|^2}{2\sigma^2} \right) \tag{2.3}$$

The Laplace RBF kernel:

$$K(x, x_i) = \exp(-\sigma \|x - x_i\|) \tag{2.4}$$

The sigmoid kernel:

$$K(x, x_i) = \tanh(\beta_0 x^T x_i + \beta_1) \tag{2.5}$$

The analysis of variance (ANOVA) RB kernel:

$$K(x, x_i) = \sum_{k=1}^{n} \exp\left(-\sigma(x^k - x_i^k)^2\right)^d \tag{2.6}$$

The linear spline kernel in one dimension:

$$K(x, x_i) = 1 + x x_i \min(x, x_i) - \frac{x + x_i}{2} \left(\min(x, x_i)^2 + \frac{(\min(x, x_i)^3}{3} \right) \tag{2.7}$$

Kernel selection is heavily dependent on the specific dataset. Currently, the most commonly used kernel functions are the Gaussian and Laplace RBF for general purpose when prior knowledge of the data is not available. Gaussian kernel avoids the sparse distribution, while the high-degree, polynomial kernel may cause the space distribution in large feature space. The polynomial kernel is widely used in image processing, while ANOVA RB is often used for regression tasks. The spline kernels are useful for continuous signal processing algorithms that involve B-spline inner products or the convolution of several spline basis functions. Thus, in this work, we adopt only the first five kernels in Equations 2.1–2.5.

A choice of appropriate kernel functions as a generic learning element has been a major problem, as classification accuracy itself heavily depends on the kernel selection. For example, Ye, Ji, and Chen [20] modified the kernel function by extending the Riemannian geometry structure induced by the kernel. Souza and de Carvalho [21] proposed selecting the hyperplanes parameters by using k-fold cross-validation and leave-one-out criteria. Ding and Dubchak [22] proposed an ad-hoc ensemble learning approach where multiclass k-nearest neighborhood classifiers were individually trained on each feature space and later combined. Damoulas and Girolami [23] proposed the use of four additional feature groups to replace the amino-acid composition. Weston et al. [24] performed feature selection for support vector machines (SVMs) by combining the feature scaling technique with the leave-one-out error bound. Chapelle et al. [25] tuned multiple parameters for two-norm SVMs by minimizing the radius margin bound or the span bound. Ong et al. [26] applied semidefinite programming to learn kernel function by hyperkernel. Lanckriet et al. [27] designed kernel matrix directly by semidefinite programming.

Multiple kernel learning (MKL) has been considered as a solution to make the kernel choice in a feasible manner. Amari and Wu [11] proposed a method of modifying a kernel function to improve the performance of a support vector machine classifier on the basis of the Riemannian geometrical structure induced by the kernel function. This idea was to enlarge the spatial resolution around the separating boundary surface by a conformal mapping such that the separability between classes can be increased in the kernel space. The experiment results showed remarkable improvement for generalization errors. Rakotomamonjy et al. [13] adopted MKL method to learn a kernel and associate predictor in supervised learning settings at the same time. This study illustrated the usefulness of MKL for some regressions based on wavelet kernels and on some model selection problems related to multiclass classification problems.

In this study, we propose a single multiclass kernel machine that is able to operate on all groups of features simultaneously and adaptively combine them. This new framework provides a new and efficient way of incorporating multiple feature characteristics without increasing the number of required classifiers. The proposed approach is based on the ability to embed each object description [28] via the kernel trick into a kernel Hilbert space. This process applies a similarity measure to every feature space. We show in this study that these similarity measures can be combined in the form of the composite kernel space. We design a new single/multiclass kernel machine that can operate composite spaces effectively by evaluating principal components of the number of kernel feature spaces. A hierarchical multiclass model enables us to learn the significance of each source/feature space, and the predictive term computed by the corresponding kernel weights may provide the regressors and the kernel parameters without resorting to ad-hoc ensemble learning, the combination of binary classifiers, or unnecessary parameter tuning.

2.2.2 Kernel Principal Component Analysis (KPCA)

KPCA uses a Mercer kernel [5] to perform a linear PCA. The gray level image of computed tomographic colonography (CTC) has been centered so that its scatter matrix of the data is given by $S = \sum_{i=1}^{n}(\Phi(x_i)\Phi(x_i)^T$. Eigenvalues λ_j and eigenvectors e_j are obtained by solving

$$\lambda_j e_j = Se_j = \sum_{i=1}^{n} \Phi(x_i)\Phi(x_i)^T e_j = \sum_{i=1}^{n} \langle e_j, \Phi(x_i)\rangle \Phi(x_i) \tag{2.8}$$

for $j = 1, \dots, n$. As Φ is not known, Equation 2.8 must be solved indirectly as proposed in the following section. Let us introduce the inner product of the transformed vectors by

$$a_{ji} = \frac{1}{\lambda_j}\langle e_j, \Phi(x_i)\rangle$$

where

$$e_j = \sum_{i=1}^{n} a_{ji}\Phi(x_i) \tag{2.9}$$

Multiplying by $\Phi(x_q)^T$ on the left, for $q = 1, \dots, n$, and substituting yields

$$\lambda_j\langle \Phi(x_q), e_j\rangle = \sum_{i=1}^{n} \langle e_j, \Phi(x_i)\rangle\langle \Phi(x_q), \Phi(x_i)\rangle \tag{2.10}$$

Substitution of Equation 2.9 into Equation 2.10 produces

$$\lambda_j \left\langle \Phi\left(x_q\right), \sum_{i=1}^{n} a_{ji}\Phi(x_i) \right\rangle = \sum_{i=1}^{n} \left(\sum_{k=1}^{n} \left\langle a_{jk}\Phi\left(x_k\right), \Phi(x_i)\right\rangle \langle \Phi(x_q), \Phi(x_i)\rangle \right) \tag{2.11}$$

which can be rewritten as $\lambda_j K a_j = K^2 a_j$, where K is an $n \times n$ Gram matrix, with the element $k_{ij} = \langle \Phi(x_i), \Phi(x_j)\rangle$ and $a_j = [a_{j1}a_{j2} \cdots a_{jn}]^T$. The latter is a dual eigenvalue problem equivalent to the problem

$$\lambda_j a_j = K a_j \tag{2.12}$$

Note that $||a_j||^2 = 1/\lambda_j$.

For example, we may choose a Gaussian kernel such as

$$k_{ij} = \langle \Phi(x_i), \Phi(x_j)\rangle = \exp\left(-\frac{1}{2\sigma^2}\left\|x_i - x_j\right\|^2\right) \tag{2.13}$$

Please note that if the image of X_n (finite sequences of inputs $x := \{x_j : j \in N_n\}$) is not centered in the Hilbert space, we need to use the centered Gram Matrix deduced by Mangasarian et al. [10] by applying the following $\hat{K}$:

$$\hat{K} = K - KT - TK + TKT \tag{2.14}$$

where K is the Gram matrix of uncentered data, and

$$T = \begin{bmatrix} \frac{1}{n} & \cdots & \frac{1}{n} \\ \cdots & \cdots & \cdots \\ \frac{1}{n} & \cdots & \frac{1}{n} \end{bmatrix}_{n \times n}$$

Let us keep the l eigenvectors associated with the l largest eigenvalues, we can reconstruct data in the mapped space: $\Phi_i' = \sum_{j=1}^{l} \langle \Phi_i, e_j \rangle e_j = \sum_{j=1}^{l} \beta_{ji} e_j$, where $\beta_{ji} = \left\langle \Phi_i, \sum_{k=1}^{n} a_{jk} \Phi_k \right\rangle = \sum_{k=1}^{n} a_{jk} k_{ik}$. For the experimental evaluation, we introduce the reconstruction square error of each data Φ_i, $i = 1, \ldots, n$, is $\text{Err}_i = \|\Phi_i - \Phi_i'\|^2 = k_{ii} - \sum_{j=1}^{\ell} \beta_{ji}^2$.

The mean square error is $\text{MErr} = (1/n) \sum_{i=1}^{n} \text{Err}_i$. Using Equation 2.12, $\beta_{ji} = \lambda_j \alpha_{ji}$. Therefore, the mean square reconstruction error is $\text{MErr} = (1/n) \sum_{i=1}^{n} \left(k_{ii} - \sum_{j=1}^{l} \lambda^2{}_j a^2{}_{ji} \right)$. As $\sum_{i=1}^{n} k_{ii} = \sum_{i=1}^{n} l_i$ and $\sum_{i=1}^{n} a_{ji}{}^2 = ||a_j||^2 = 1/\lambda_j$, $\text{MErr} = \frac{1}{n} \sum_{i=\ell+1}^{n} \lambda_i$.

KPCA algorithm contains an eigenvalue problem of rank n, so the computational complexity of KPCA is $O(n^3)$. In addition, each resulting eigenvector is represented as a linear combination of n terms; the l features depend on n image vectors of X_n. Thus, all data contained in X_n must be retained, which is computationally cumbersome and unacceptable for our applications.

KPCA algorithm [5]

Step 1 Calculate the Gram matrix, which contains the inner products between pairs of image vectors.

Step 2 Use $\lambda_j a_j = K a_j$ to obtain the coefficient vectors a_j for $j = 1, \ldots, n$.

Step 3 The projection of $x \in R^d$ along the jth eigenvector is

$$\langle e_j, \Phi(x) \rangle = \sum_{i=1}^{n} a_{ji} \langle \Phi(x_i), \Phi(x) \rangle = \sum_{i=1}^{n} a_{ji} k(x, x_i)$$

2.2.3 Accelerated Kernel Feature Analysis (AKFA)

AKFA [29] is the method that we have proposed to improve the efficiency and accuracy of SKFA [10] by Mangasarian et al. SKFA improves the computational costs of

KPCA, associated with both time complexity and data retention requirements. SKFA was introduced in [10] and is summarized in the following three steps:

SKFA algorithm [10]

Step 1 Compute the matrix $k_{ij} := k(x_i, x_j)$, it costs $O(d.n_2)$ operations, where d is the dimensionality of input space X.

Step 2 Initialize $a_{01}, \ldots, a_{0m} = 1$, and $idx(\cdot)$ as the empty list, these are the initial scaling for the directions of projection. It costs $O(m)$.

Step 3 For $i = 1$ to I repeat: (I represent the number of features to be extracted)

1. Compute the Q values based on $\Phi(x_1), \ldots, \Phi(x_m)$ for all directions $\Phi_{i1}, \ldots, \Phi_{i,m-i+1}$. It can be got by $\langle \Phi_j^i, \Phi(x_l) \rangle = a_{0j} k_{jl} + \sum_{t=1}^{i-1} a_{tl} k_{idx(t),l}$. It costs $O(i.m^2)$ steps since we need i operations per dot product. Compute the Q value for each direction Φ_j^i.
2. Perform a maximum search over all Q values ($O(m)$) and pick the corresponding Φ_j^i, this is the ith principal direction v_i, and store the corresponding coefficients $a_{1j}, \ldots, a_{ij}$, set $idx(i) = j$.
3. Compute the new search space to perform orthogonalization by $\Phi_j^{i+1} := \Phi_j^i - v_i \langle \Phi_j^i, v_i \rangle / \|v_i\|^2$. All coefficients have to be stored into a_{ij}. All entries $\Phi(x_j)$, concerning are sorted into a_{jl} with $1 \leq l \leq m$, respectively. The other coefficients are assigned to atl with $1 \leq t \leq i$ and $1 \leq l \leq m$.

AKFA [5] has been proposed by the author in an attempt to achieve further improvements: (i) saves computation time by iteratively updating the Gram matrix, (ii) normalizes the images with the l_2 constraint before the l_1 constraint is applied, and (iii) optionally discards data that falls below a magnitude threshold δ during updates.

To achieve the computation efficiency described in (i), instead of extracting features directly from the original mapped space, AKFA extracts the ith feature on the basis of the ith updated Gram matrix K^i, where each element is $k_{jk}{}^i = \langle \Phi_j{}^i, \Phi_k{}^i \rangle$.

The second improvement described previously in (ii) is to revise the l_1 constraint. SKFA treats each individual sample data as a possible direction and computes the projection variances with all data. As SKFA includes its length in its projection variance calculation, it is biased to select vectors with larger magnitude. We are ultimately looking for a direction with unit length, and when we choose an image vector as a possible direction, we ignore the length and only consider its direction for the improved accuracy of the features.

The third improvement in (iii) is to discard negligible data and thereby eliminate unnecessary computations.

AKFA is described in the following three steps and showed the improvements (i)–(iii) [29]. The vector Φ_i' represents the reconstructed new data based on AKFA, and it can be calculated indirectly using the kernel trick: $\Phi_i = \sum_{j=1}^{\ell} \langle \Phi_i, v_j \rangle v_j =$

$\Phi_\ell C_\ell C_\ell^T K_i$, where $K_i = [k_i; idx(1)\ \ k_i; idx(2); \ldots ; k_i; idx(l)]^T$. Then the reconstruction error of new data Φ_i, $i = n+1, \ldots, n+m$, is represented as

$$\text{Err}_i = \|\Phi_i - \Phi_i'\|^2 = k_{ii} - K_i^T C_l C_l^T K_i$$

AKFA algorithm [29]

Step 1 Compute the $n \times n$ Gram matrix $k_{ij} = k(x_i, x_j)$, where n is the number of input vectors. This part requires $O(n^2)$ operations.

Step 2 Let l denote the number of features to be extracted. Initialize the $l \times l$ coefficient matrix $\mathbf{C}$ to $\mathbf{0}$, and $idx(\cdot)$ as an empty list which will ultimately store the indices of the selected image vectors, and $\mathbf{C}_{(i-1)}$ is an upper-triangle coefficient matrix. Let us define $\Phi'^i{}_{idx(i)} = \Phi_{idx(i)} \quad - \sum_{t=1}^{i-1} \langle \Phi_{idx(i)}, v_t \rangle v_t$. Initialize the threshold value $\delta = 0$ for the reconstruction error. The overall cost is $O(l^2)$.

Step 3 For $i = 1$ to l repeat:

1. Using the ith updated K^i matrix, extract the ith feature. If $k_{jj}{}^i < \delta$, the predetermined $\delta > 0$. It is a threshold that determines the number of features we selected. Then discard jth column and jth row vector without calculating the projection variance. Use $idx(i)$ to store the index. This step requires $O(n^2)$ operations.
2. Update the coefficient matrix by using $\mathbf{C}_{i,i} = 1/\sqrt{k^i_{idx(i),idx(i)}}$ and $\mathbf{C}_{1:(i-1),i} = -\mathbf{C}_{i,i}\mathbf{C}_{(\mathbf{i}-1)}\mathbf{C}^{\mathbf{T}}_{(\mathbf{i}-1)}\mathbf{K}_{\mathbf{idx(i)}\cdot}$, which requires $O(i^2)$ operations.
3. Obtain $\mathbf{K}^{i+1}$, an updated Gram matrix. Neglect all rows and columns containing diagonal elements less than δ. This step requires $O(n^2)$ operations. The total computational complexity is increased to $O(ln^2)$ when no data is being truncated during updating in the AKFA.

AKFA algorithm also contains an eigenvalue problem of rank n, so the computational complexity of AKFA requires $O(n^2)$ operations in Step 1, Step 2 is $O(l^2)$. Step 3 requires 1 for $O(n^2)$, 2 for $O(i^2)$, and 3 for $O(n^2)$. The total computational complexity is increased to $O(ln^2)$ when no data is being truncated during updating in the AKFA.

2.2.4 Comparison of the Relevant Kernel Methods

Multiple kernel adoption and combination methods are derived from the principle of empirical risk minimization, which performs well in most applications. Actually, to access the expected risk, there is an increasing amount of literature focusing on the theoretical approximation error bounds with respect to the kernel selection problem, for example, empirical risk minimization, structural risk minimization, approximation error, approximation error, span bound, Jaakkola–Haussler bound,

Table 2.1 Overview of method comparison for parameters tuning

Method	Principle	(Dis)Advantage
Empirical risk minimization [30]	Averaging the loss function on the training set for unknown distribution	High variance, poor generalization, overfitting
Structural risk minimization [30]	Incorporating a regularization penalty into the optimization	Low bias, high variance, prevent overfitting
Approximation error [31]	Featuring diameter of the smallest sphere containing the training points	Expensive computation
Span bound [32]	Applying a gradient descent method through learning the distribution of kernel functions	Optimal approximation of an upper bound of the prediction risk
Jaakkola–Haussler bound [33]	Computing the leave-one-out error and the inequality	Loose approximations for bounds
Radius-margin bound [34]	Calculating the gradient of bound value with the scaling factor	Optimal parameters depending on the performance measure
Kernel Linear Discriminant Analysis [35]	Extending an online LDA via a kernel trick	Complicated characteristics of kernel discriminant analysis

radius-margin bound, and kernel linear discriminant analysis. Table 2.1 lists some comparative methods among the multiple kernel methods.

2.3 PRINCIPAL COMPOSITE KERNEL FEATURE ANALYSIS (PC-KFA)

2.3.1 Kernel Selections

For kernel-based learning algorithms, the key challenge lies in the selection of kernel and regularization parameters. Many researchers have identified this problem and thus have tried to solve it. However, the few existing solutions lack effectiveness, and thus this problem is still under development or regarded as an open problem. To this end, we are developing a new framework of kernel adaptation. Our method exploits the idea presented in Refs. [36, 37], by exploring data-dependent kernel methodology as follows:

Let $\{x_i, y_i\}(i = 1, 2, \ldots, n)$ be n d-dimensional training samples of the given data, where $y_i = \{+1, -1\}$ represents the class labels of the samples. We develop a data-dependent kernel to capture the relationship among the data in this classification task by adopting the idea of "conformal mapping" [37]. To adopt this conformal transformation technique, this data-dependent composite kernel for $r = 1, 2, 3, 4, 5$ can be formulated as

$$k_r(x_i, x_j) = q_r(x_i)q_r(x_j)p_r(x_i, x_j) \tag{2.15}$$

where $p_r(x_i, x_j)$ is one kernel among five chosen kernels and $q(\cdot)$, the factor function, takes the following form for $r = 1, 2, 3, 4, 5$:

$$q_r(x_i) = \alpha_{r0} + \sum_{m=1}^{n} \alpha_{rm} k_0(x_i, x_m) \tag{2.16}$$

where $k_0(x_i, x_m) = \exp(-||x_i - x_m||^2 / 2\sigma^2)$, and α_{rm} is the combination coefficient. Let us denote the vectors $\{q_r(x_1), q_r(x_2), \dots, q_r(x_n)\}^T$ and $\{a_0, a_1, a_n\}_r^T$ by q_r and α_{rm} ($r = 1, 2, 3, 4, 5$), respectively, where we have $q_r = K_0\alpha_r$, where K_0 is a $n \times (n+1)$ matrix given by

$$K_0 = \begin{bmatrix} 1 & k_0(x_1, x_1) & \cdots & k_0(x_1, x_n) \\ 1 & k_0(x_2, x_1) & \cdots & k_0(x_2, x_n) \\ \vdots & \vdots & \cdots & \vdots \\ 1 & k_0(x_n, x_1) & \cdots & k_0(x_n, x_n) \end{bmatrix} \tag{2.17}$$

Let the kernel matrices corresponding to $k(x_i, x_j), p_1(x_i, x_j)$ and $p_2(x_i, x_j)$ be K, P_1 and P_2 respectively. We can express data-dependent kernel K as

$$K^* = [q_r(x_i)q_r(x_j)p_r(x_i, x_j)]_{n \times n} \tag{2.18}$$

Defining Q_i as the diagonal matrix of elements $\{q_i(x_1),\ q_i(x_2), \dots, q_i(x_{xn})\}$, we can express Equation 2.18 as the matrix form:

$$K_r = Q_r P_r Q_r \tag{2.19}$$

This kernel model was first introduced in [20] and called "conformal transformation of a kernel." We now perform kernel optimization on the basis of the method to find the appropriate kernels for the data set.

The optimization of the data-dependent kernel in Equation 2.19 is to set the value of combination coefficient vector α_r so that the class separability of the training data in mapped feature space is maximized. For this purpose, Fisher scalar is adopted as the objective function of our kernel optimization. Fisher scalar measures the class separability of the training data in the mapped feature space and is formulated as

$$J = \frac{\mathrm{tr}(S_{br})}{\mathrm{tr}(S_{wr})} \tag{2.20}$$

where S_{b1}, S_{b2} represents the "between-class scatter matrices" and S_{w1}, S_{w2} are the "within-class scatter matrices." Suppose that the training data are grouped according to their class labels, that is, the first n_1 data belong to one class and the remaining n_2 data belong to the other class ($n_1 + n_2 = n$). Then, the basic kernel matrix P_i can be partitioned as

$$P_r = \begin{pmatrix} P^r_{11} & P^r_{12} \\ P^r_{21} & P^r_{22} \end{pmatrix} \tag{2.21}$$

where the sizes of the submatrices $P^r_{11}, P^r_{12}, P^r_{21}, P^r_{22}$, $r = 1, 2, 3, 4, 5$, are $n_1 \times n_1, n_1 \times n_2, n_2 \times n_1, n_2 \times n_2$, respectively.

A close relation between the class separability measure J and the kernel matrices has been established as

$$J(\alpha_r) = \frac{\alpha_r^T M_{0r} \alpha_r}{\alpha_r{}^T N_{0r} \alpha_r} \tag{2.22}$$

where

$$M_{0r} = K_0{}^T B_{0r} K_0 \quad \text{and} \quad N_{0r} = K_0{}^T W_{0r} K_0 \tag{2.23}$$

And for $r = 1, 2, 3, 4, 5$,

$$B_{0r} = \begin{pmatrix} \frac{1}{n_1} P^r_{11} & 0 \\ 0 & \frac{1}{n_2} P^r_{22} \end{pmatrix} - \frac{1}{n} P_r \tag{2.24}$$

$$W_{0r} = \text{diag}(p^r_{11}, p^r_{22}, \ldots, p^r_{nn}) - \begin{pmatrix} \frac{1}{n_1} P^r_{11} & 0 \\ 0 & \frac{1}{n_2} P^r_{22} \end{pmatrix}$$

To maximize $J(\alpha_r)$ in Equation 2.22, the standard gradient approach is followed. If matrix N_{0i} is nonsingular, the optimal α_i that maximizes $J_i(\alpha_i)$ is the eigenvector corresponding to the maximum eigenvalue of the system, we will drive Equation 2.25 as taking the derivatives.

$$M_{0r} \alpha_r = \lambda_r N_{0r} \alpha_r \tag{2.25}$$

The criterion for selecting the best kernel function is to find the kernel that produces the largest eigenvalue from Equation 2.25, that is,

$$\lambda_r^* = \arg\max(N_r{}^{-1} M_r) \tag{2.26}$$

The idea behind it is to choose the maximum eigenvector α_i corresponding to the maximum eigenvalue that can maximize the $J_i(\alpha_i)$ that will result in the optimum solution. We find the maximum eigenvalues for all possible kernel functions and arrange them in descending order to choose the most optimum kernels, such as

$$\lambda_1^* > \lambda_3^* > \lambda_4^* > \lambda_2^* > \lambda_5^* \tag{2.27}$$

We choose the kernels corresponding to the largest eigenvalues $\lambda_1{}^*$ and forming composite kernels corresponding to $\{\lambda_1{}^*, \lambda_3{}^*, \ldots\}$ as follows:

Kernel Selection Algorithm

Step 1 Group the data according to their class labels. Calculate P_r, K_1 first and then B_{0r}, W_{0r} through which we can calculate M_{0r} and N_{0r} for $r = 1, 2, 3, 4, 5$.

Step 2 Calculate the eigenvalue α_r^* corresponding to maximum eigenvector $\lambda_r^* = \arg\max_{\lambda}(N^{-1}M)$.

Step 3 Arrange the eigenvalues in the descending order of magnitude.

Step 4 Choose the kernels corresponding to most dominant eigenvalues.
Step 5 Calculate $q_r = K_1 a_r{}^*$.
Step 6 Calculate Q_r and then compute $Q_r P_r Q_r$ for the most dominant kernels.

2.3.2 Kernel Combinatory Optimization

In this section, we propose a principal composite kernel function that is defined as the weighted sum of the set of different optimized kernel functions [15, 16]. To obtain an optimum kernel process, we define the following composite kernel as

$$K_{comp}(\rho) = \sum_{i=1}^{p} \rho_i Q_i P_i Q_i \tag{2.28}$$

where ρ is the constant scalar value of the composite coefficient and p is the number of kernels we intend to combine. Through this approach, the relative contribution of both kernels to the model can be varied over the input space. We note that in Equation 2.28, instead of using K_r as a kernel matrix, we use K_{comp} as a composite Kernel Matrix. According to [38], K_{comp} satisfies the Mercers condition. We use linear combination of individual kernels to yield an optimal composite kernel using the concept of kernel alignment, "conformal transformation of a kernel." The empirical alignment between kernel k_1 and kernel k_2 with respect to the training set S is the following quantity metric:

$$A(k_1, k_2) = \frac{\langle K_1, K_2 \rangle_F}{\|K_1\|_F \|K_2\|_F} \tag{2.29}$$

where K_i is the kernel matrix for the training set S using kernel function k_i, and $\|K_i\|_F = \sqrt{\langle K_i, K_i \rangle_F}$, $\langle K_i, K_j \rangle_F$ is the Frobenius inner product between K_i and K_j. $S = \{(x_i, y_i) \,|\, x_i \in X, y_i \in \{+1, -1\}, i = 1, 2, \ldots, n\}$, X is the input space, y is the target vector. Let $K_2 = yy'$, then the empirical alignment between kernel k and target vector y is

$$A(k, yy') = \frac{\langle K, yy' \rangle_F}{\|K\|_F \|yy'\|_F} = \frac{y'Ky}{n\|K\|_F} \tag{2.30}$$

It has been shown that if a kernel is well aligned with the target information, there exists a separation of the data with a low bound on the generalization error. Thus, we can optimize the kernel alignment on the basis of training set information to improve the generalization performance of the test set. Let us consider the combination of kernel functions as follows:

$$k(\rho) = \sum_{i=1}^{p} \rho_i k_i \tag{2.31}$$

where individual kernels $k_i, i = 1, 2, \dots, p$ are known in advance. Our purpose is to tune ρ to maximize $A(\rho, k, yy')$ the empirical alignment between $k(\rho)$ and the target vector y. Hence, we have

$$\rho = \arg\max(A(\rho, k, yy')) \tag{2.32}$$

$$= \arg\max\left(\frac{\left\langle \sum_i \rho_i K_i, yy' \right\rangle}{n\sqrt{\left\langle \sum_i \rho_i K_i \right\rangle, \left\langle \sum_j \rho_j K_j \right\rangle}}\right) = \arg\max\left(\frac{\sum_i \rho_i \left\langle K_i, yy' \right\rangle}{n\sqrt{\sum_{i,j} \rho_i, \rho_j \langle K_i, K_j \rangle}}\right)$$

$$= \arg\max\left(\frac{\left(\sum_i \rho_i u_i\right)^2}{n^2 \sum_{i,j} \rho_i \rho_j v_{ij}}\right) = \arg\max\left(\frac{1}{n^2} \cdot \frac{\rho^T U \rho}{\rho^T V \rho}\right) \tag{2.33}$$

where $u_i = \sqrt{\langle K_i, yy^T \rangle}, v_{ij} = \sqrt{\langle K_i, K_j \rangle}, U_{ij} = u_i u_j, V_{ij} = v_i v_j$, and $\rho = (\sqrt{\rho_1}, \sqrt{\rho_2}, \dots, \sqrt{\rho_p})$.

Let the generalized Raleigh coefficient be

$$J(\rho) = \frac{\rho^T U \rho}{\rho^T V \rho} \tag{2.34}$$

Therefore, we can obtain the value of $\hat{\rho}$ by solving the generalized eigenvalue problem

$$U\rho = \delta V \rho \tag{2.35}$$

where δ denotes the eigenvalues.

PC-KFA Algorithm

Step 1 Compute optimum parameter $\hat{\rho}$ in $U\rho = \delta V\rho$.

Step 2 Implement $K_{comp}(\rho)$ for optimum parameter $\hat{\rho}$.

Step 3 Build the model with $K_{comp}(\rho)$ using all training data.

Step 4 Test the completed model on the test set.

PC-KFA algorithm contains an eigenvalue problem of rank n, so the computational complexity of PC-KFA requires $O(n^2)$ operations in Step 1, Step 2 is n. Step 3 requires n operations. Step 4 requires n operations. The total computational complexity is increased to $O(n^2)$.

2.4 EXPERIMENTAL ANALYSIS

2.4.1 Cancer Image Datasets

Colon Cancer This dataset consisted of true-positive (TP) and false-positive (FP) detections obtained from our previously developed CAD scheme for the detection of polyps [39], when it was applied to a CTC image database. This database contained 146 patients who underwent a bowel preparation regimen with a standard precolonoscopy bowel-cleansing method. Each patient was scanned in both supine and prone positions, resulting in a total of 292 computed tomography (CT) datasets. In the scanning, helical single-slice or multislice CT scanners were used, with collimations of 1.25–5.0 mm, reconstruction intervals of 1.0–5.0 mm, X-ray tube currents of 50–260 mA, and voltages of 120–140 kVp. In-plane voxel sizes were 0.51–0.94 mm, and the CT image matrix size was 512×512. Out of 146 patients, there were 108 normal cases and 38 abnormal cases with a total of 39 colonoscopy-confirmed polyps larger than 6 mm.

The CAD scheme was applied to the entire cases, and it generated a segmented region for each of its detection (a candidate of polyp). A volume of interest (VOI) of size $64 \times 64 \times 64$ voxels was placed at the center of mass of each candidate for encompassing its entire region; then, it was resampled to $12 \times 12 \times 12$ voxels. Resulting VOIs of 39 TP and 149 FP detections from the CAD scheme made up the colon cancer dataset 1.

Additional CTC image databases with a similar cohort of patients were collected from three different hospitals in the United States. The VOIs obtained from these databases were resampled to $16 \times 16 \times 16$ voxels. We named the resulting datasets as colon cancer datasets 2, 3, and 4. Tables 2.2–2.5 list the distribution of the training and testing VOIs in the colon cancer datasets 1, 2, 3, and 4, respectively.

Breast Cancer We extended our own colon cancer datasets into other cancer-relevant datasets. This dataset is available at http://www.ncbi.nlm.nih.gov/geo/

Table 2.2 Colon cancer dataset 1 (low resolution)

		Portion, %	Data portion	Data size
Training set	TP	80.0	31	148
	FP	78.3	117	
Testing set	TP	20.0	8	40
	FP	21.7	32	

Table 2.3 Colon cancer dataset 2 (U. Chicago)

		Portion, %	Data portion	Data size
Training set	TP	80.0	16	766
	FP	70.0	750	
Testing set	TP	20.0	16	316
	FP	30.0	300	

Table 2.4 Colon cancer dataset 3 (BID)

		Portion, %	Data portion	Data size
Training set	TP	80.0	22	1012
	FP	70.0	990	
Testing set	TP	20.0	6	431
	FP	30.0	425	

Table 2.5 Colon cancer dataset 4 (NorthWestern U.)

		Portion, %	Data portion	Data size
Training set	TP	80.0	17	1817
	FP	60.0	1800	
Testing set	TP	20.0	4	1204
	FP	40.0	1200	

Table 2.6 Breast cancer dataset

		Portion, %	Data portion	Data size
Training set	TP	80.0	51	126
	FP	60.0	75	
Testing set	TP	20.0	13	63
	FP	40.0	50	

query/acc.cgi?acc=GSE2990. This dataset contains data on 189 women, 64 of which were treated with tamoxifen, with primary operable invasive breast cancer, with each feature dimension of 22,283. More information on this dataset can be found in [40] (Table 2.6).

Lung Cancer This dataset is available at http://www.broadinstitute.org/cgi-in/cancer/datasets.cgi. It contains 160 tissue samples, 139 of which are of class "0" and the remaining are of class "2." Each sample is represented by the expression levels of 1000 genes for each feature dimension (Table 2.7).

Lymphoma This dataset is available at http://www.broad.mit.edu/mpr/lymphoma. It contains 77 tissue samples, 58 of which are diffuse large B-cell lymphomas (DLBCL) and the remainder is follicular lymphomas (FL), with each feature dimension of 7129. Detailed information about this dataset can be found in [41] (Table 2.8).

Prostate Cancer This dataset is collected from http://www.ncbi.nlm.nih.gov/geo/query/acc.cgi?acc=GSE6919. It contains prostate cancer data collected from 308 patients, 75 of which have metastatic prostate tumor and the rest of the cases were normal, with each feature dimension of 12,553. More information on this data set can be found in [42, 43] (Table 2.9).

Table 2.7 Lung cancer dataset

		Portion, %	Data portion	Data size
Training set	TP	70.0	15	126
	FP	80.0	111	
Testing set	TP	30.0	6	34
	FP	20.0	28	

Table 2.8 Lymphoma dataset

		Portion, %	Data portion	Data size
Training set	TP	85.0	17	62
	FP	78.0	45	
Testing set	TP	15.0	3	14
	FP	22.0	13	

Table 2.9 Prostate cancer dataset

		Portion, %	Data portion	Data size
Training set	TP	85.0	64	250
	FP	80.0	186	
Testing set	TP	15.0	11	58
	FP	20.0	47	

2.4.2 Kernel Selection

We first evaluate the performance on the kernel selection according to the method proposed in Section 2.3.1, regarding how to select the kernel function that will best fit the data. The larger the eigenvalue is, the greater the class separability measure J in Equation 2.22 is to be expected. Table 2.10 shows the calculation of the algorithm for all the datasets mentioned to determine the eigenvalues of all five kernels. Specifically, we have set the parameters such as d, offset, β_0, β_1, and σ of each kernel in Equations 2.1–2.5), and computed their eigenvalues λ for all the eight datasets from Tables 2.2–2.9. After arranging the eigenvalues for each dataset in descending order, we selected the kernel corresponding to the largest eigenvalue as the optimum kernel.

The largest eigenvalue for each data set is highlighted in Table 2.10. After evaluating the quantitative eigenvalues for all the eight datasets, we observed that the RBF kernel gives the maximum eigenvalue among all the five kernels. This means that RBF kernel produced the dominant results compared to all other four kernels. For four datasets, colon cancer datasets 2, 3, 4: lymphoma cancer dataset, the Polynomial kernel produced the second largest eigenvalue. Linear kernel gave the second largest eigenvalue for colon cancer dataset 1 and lung cancer dataset, where as the Laplace kernel produced the second largest eigenvalue for the breast cancer dataset.

As shown in Table 2.10, the Gaussian GBF kernel showed largest eigenvalues for the all eight datasets. The performance of the selected Gaussian GBF was compared

Table 2.10 Eigenvalues λ of five kernel functions (Eqs. 2.1-2.5) and their parameters selected

Cancer datasets	Linear	Polynominal	Gaussian RBF	Laplace RBF	Sigmoid
Colon1	13.28	11.54	**16.82**	7.87	3.02
		$d = 1$, Offset = 1	$\sigma = 4.00$	$\sigma = 0.1$	$\beta_0 = 2, \beta_1 = 1.7$
Colon2	75.43	84.07	**139.96**	40.37	64.5
		$d = 1$, Offset = 4	$\sigma = 5.65$	$\sigma = 1.5$	$\beta_0 = 1,\ \beta_1 = 2.5$
Colon3	100.72	106.52	**137.74**	80.67	53.2
		$d = 1$, Offset = 1	$\sigma = 4.47$	$\sigma = 1.5$	$\beta_0 = 2,\ \beta_1 = 3$
Colon4	148.69	166.44	**192.14**	34.99	142.3
		$d = 1$, Offset = 1	$\sigma = 4.58$	$\sigma = 1.5$	$\beta_0 = 2,\ \beta_1 = 2$
Breast	22.85	20.43	**64.38**	56.85	23.2
		$d = 1.2$, Offset = 1	$\sigma = 4.47$	$\sigma = 3.0$	$\beta_0 = 0.75,\ \beta_1 = 1$
Lung	36.72	47.49	**54.60**	38.74	29.2
		$d = 1.2$, Offset = 4	$\sigma = 3.87$	$\sigma = 2.4$	$\beta_0 = 4,\ \beta_1 = 2.5$
Lymphoma	19.71	37.50	**42.13**	35.37	23.6
		$d = 1.5$, Offset = 2	$\sigma = 2.82$	$\sigma = 2.0$	$\beta_0 = 1.5,\ \beta_1 = 2$
Prostate	50.93	48.82	**53.98**	40.33	43.1
		$d = 1$, Offset = 1	$\sigma = 4.47$	$\sigma = 1.5$	$\beta_0 = 0.5,\ \beta_1 = 0.5$

Table 2.11 Mean square reconstruction error of KPCA, SKFA, and AKFA with the selected kernel function

Cancer datasets	Selected kernel function	Eigenspace dimension	KPCA error, %	SKFA error, %	AKFA error, %
Colon1	RBF	75	6.86	11.56	10.74
Colon2	RBF	100	27.08	18.41	17.00
Colon3	RBF	100	14.30	22.29	20.59
Colon4	RBF	100	12.48	19.66	18.14
Breast	RBF	55	6.05	2.10	10.10
Lung	RBF	50	1.53	2.55	7.30
Lymphoma	RBF	20	3.27	7.2	3.87
Prostate	RBF	80	10.33	11.2	13.83

to the other single kernel function in the reconstruction error value. As a further experiment, the reconstruction error results have been evaluated for KPCA using $\mathrm{MErr} = (1/n)\sum_{i=l+1}^{n} \lambda_i$, and for AKFA and SKFA using $\mathrm{Err}_i = ||\Phi_i - \Phi_i'||^2 = k_{ii} - K_i{}^T C_l C^T{}_l K_i$ with the optimum kernel (RBF) selected from Table 2.10. We listed up the selected kernel, dimensions of the eigenspace (chosen empirically) and the reconstruction errors of KPCA, SKFA, and AKFA for all the datasets shown in Table 2.11.

Table 2.11 shows that RBF, the single kernel selected, has a relatively small reconstruction error, from 3.27% to up to 14.30% in KPCA. The reconstruction error of KPCA is less than that of the reconstruction error of AKFA, from 0.6% to up to 6.29%. The difference in the reconstruction error between KPCA and AKFA increased as the

Table 2.12 Mean square reconstruction error of KPCA with other 4 kernel functions

Cancer datasets	Linear kernel function, %	Polynominal kernel function, %	Laplace RBF kernel, %	Sigmoid kernel function, %
Colon1	1,739	1,739	46.43	238.6
Colon2	12,133	33,170	90.05	291.1
Colon3	4,276	4,276	38.41	294.6
Colon4	1,972	1,972	26.28	228.6
Breast	477.6	2,061	49.63	465.3
Lung	1,009	5,702	59.51	464.8
Lymphoma	362.5	362.5	63.04	228.5
Prostate	849.8	849.8	67.44	159.8

size of the datasets increased. This could be due to the heterogeneous nature of the datasets. The Lymphoma dataset produced the least mean square error, whereas the colon cancer dataset 3 produced the largest mean square error for both KPCA and AKFA.

Table 2.12 shows that the other four kernel functions have much more error than Gaissian RBF shown in Table 2.11. The difference between Tables 2.11 and 2.12 is more than four times larger reconstruction error, and sometimes 20 times when the other four kernel functions are applied.

2.4.3 Kernel Combination and Reconstruction

After selecting the number of kernels, we select the first p kernels that produced the p largest eigenvalues in Table 2.10, and combine them according to the method proposed in Section 2.3.2 to yield lesser reconstruction error. Table 2.13 shows the coefficients calculated for the linear combination of kernels. After obtaining the linear coefficients according to Equation 2.35, we combine the kernels according to Equation 2.28 to generate the composite Kernel Matrix $K_{comp}(\rho)$. Table 2.14 shows the reconstruction error results for both KPCA and AKFA along with the composite kernel $K_{comp}(\rho)$.

Table 2.13 Linear combination $\hat{\rho}$ for selected two kernel functions

Cancer datasets	Two selected kernels	Linear combination of kernels
Colon1	RBF + Linear	$\hat{\rho}_1 = 0.9852, \hat{\rho}_2 = 0.1527$
Colon2	RBF + Polynomial	$\hat{\rho}_1 = 0.6720, \hat{\rho}_2 = 0.1582$
Colon3	RBF + Polynomial	$\hat{\rho}_1 = 0.9920, \hat{\rho}_2 = 0.1204$
Colon4	RBF + Polynomial	$\hat{\rho}_1 = 0.9775, \hat{\rho}_2 = 0.1375$
Breast	RBF + Laplace	$\hat{\rho}_1 = 0.8573, \hat{\rho}_2 = 0.1386$
Lung	RBF + Linear	$\hat{\rho}_1 = 0.9793, \hat{\rho}_2 = 0.1261$
Lymphoma	RBF + Polynomial	$\hat{\rho}_1 = 0.9903, \hat{\rho}_2 = 0.2082$
Prostate	RBF + Linear	$\hat{\rho}_1 = 0.9756, \hat{\rho}_2 = 0.1219$

Table 2.14 Mean square reconstruction error with kernel combinatory optimization

Cancer datasets	Eigenspace dimension	KPCA error, %	SKFA error, %	AKFA error, %	PC-KFA %
Colon1	75	4.20	6.34	4.30	4.18
Colon2	100	5.53	7.23	5.20	5.17
Colon3	100	5.23	7.70	7.29	5.21
Colon4	100	10.50	15.17	14.16	10.48
Breast	55	2.88	3.47	6.56	2.78
Lung	50	2.43	3.71	3.67	2.44
Lymphoma	20	2.01	3.11	4.44	2.12
Prostate	80	1.34	2.23	1.06	1.28

The reconstruction error using two composite kernel functions shown in Table 2.14 is smaller than the reconstruction error in the single kernel function RBF in Table 2.11. This would lead us to claim that all eight datasets from Table 2.14 made it evident that the reconstruction ability of kernel optimized KPCA and AKFA gives enhanced performance to that of single kernel KPCA and AKFA. The specific improvement in the reconstruction error performance is greater by up to 4.27% in the case of KPCA, and by up to 5.84% and 6.12% in the cases of AKFA and SKFA, respectively. The best improvement of the error performance is observed in PC-KFA by 4.21%. This improvement in reconstruction of all datasets is validated using PC-KFA. This successfully shows that the composite kernel produces only a small reconstruction error.

2.4.4 Kernel Combination and Classification

In order to analyze how feature extraction methods affect classification performance of polyp candidates, we used the k-nearest neighborhood classifier on the image vectors in the reduced eigenspace. We evaluated the performance of classifiers, by applying KPCA and AKFA with both selected as single and composite kernels for all the eight datasets. Six nearest neighbors were used for the classification purpose. The classification accuracy was calculated as $(TP + TN)/(TP + TN + FN + FP)$. The results of classification accuracy showed very high values as shown in Table 2.15.

The results from Table 2.15 indicate that the classification accuracy of the composite kernel is better than that of the single kernel for both KPCA and AKFA in colon cancer dataset 1, breast cancer, lung cancer, lymphoma, and prostate cancer, whereas in the case of colon cancer datasets 2, 3, and 4, because of the huge size of the data, the classification accuracy is very similar between single and composite kernels. From this quantitative characteristic among the entire eight datasets, we can evaluate that the composite kernel improved the classification performance, and with single and composite kernel cases the classification performance of AKFA is equally good as that of KPCA, from 95.28% up to 91.997%. The best classification performance has been shown in PC-KFA, up to 96.79%.

Table 2.15 Classification accuracy using six nearest neighborhoods for single-kernel and two-composite-kernels with KPCA, SKFA, AKFA, and PC-KFA

Cancer datasets	KPCA single	KPCA composite	SKFA single	SKFA composite	AKFA single	AKFA composite	PC-KFA
Colon1	97.50	97.50	92.50	97.50	95.00	95.00	97.61
Colon2	86.02	86.02	86.02	86.02	85.48	86.02	86.13
Colon3	98.61	98.61	98.61	98.61	98.61	98.61	98.82
Colon4	99.67	99.67	99.67	99.67	99.67	99.67	99.70
Breast	87.50	98.41	96.81	98.41	95.21	96.83	98.55
Lung	91.18	97.06	94.12	94.12	91.18	94.12	97.14
Lymphoma	87.50	93.75	93.75	93.75	97.50	93.75	97.83
Prostate	87.96	94.83	91.38	98.28	89.66	98.28	98.56

Table 2.16 Overall classification comparison among other multiple kernel methods

Datasets	Regularized kernel discriminant analysis (RKDA) [20]	L2 regularization [44]*	Generality multiple kernel learning (GMKL) [45]	Proposed PC-KFA
Heart	73.21	0.17	NA	81.21
Cancer	95.64	NA	NA	95.84
Breast	NA	0.03	NA	84.32
Ionosphere	87.67	0.08	94.4	95.11
Sonar	76.52	0.16	82.3	84.02
Parkinsons	NA	*NA*	92.7	93.17
Musk	NA	*NA*	93.6	93.87
Wpbc	NA	*NA*	80.0	80.56

Abbreviation: NA, not available.
*Misclassification rate

2.4.5 Comparisons of Other Composite Kernel Learning Studies

In this section, we make an experimental comparison of the proposed PC-KFA with other popular MKL techniques. Such as regularized kernel discriminant analysis (RKDA) for MKL [20], L2 regulation learning [44], and generality multiple kernel learning [45] in Table 2.16, as follows:

To evaluate algorithms [20, 44, 45], the eight datasets are used in the binary-class case from the UCI Machine Learning Repository [27, 46]. L2 Regulation Learning [44] showed miss-classification ratio, which may not be equally comparative to the other three methods. The proposed PC-KFA overperformed these representative approaches. For example, PC-KFA for Lung was 97.14%, not as good as the performance of Single or two composite-kernels (SMKL), but better than RKDA for MKL and Generality multiple kernel learning (GMKL). The classification accuracy of RKDA for MKL in Dataset 4 and Prostate is better than GMKL. This result indicates PC-KFA is very competitive to the well-known classifiers for multiple datasets.

Table 2.17 PC-KFA Computation time for kernel selection and operation with KPCA, SKFA, AKFA, and PC-KFA

Cancer datasets	KPCA, s	SKFA, s	AKFA, s	PC-KFA, s
Colon1	0.266	3.39	0.218	6.12
Colon2	2.891	5.835	1.875	10.01
Colon3	6.83	16.44	3.30	21.25
Colon4	31.92	47.17	11.23	93.41
Breast	0.266	0.717	0.219	1.37
Lung	0.125	0.709	0.0625	1.31
Lymphoma	0.0781	0.125	0.0469	0.27
Prostate	1.703	4.717	1.109	9.31

2.4.6 Computation Time

We finally evaluate the computational efficiency of the proposed PC-KFA method by comparing its run time with KPCA and AKFA for all eight datasets as shown in Table 2.17. The algorithms have been implemented in MATLAB® R2007b using the Statistical Pattern Recognition Toolbox for the Gram matrix calculation and kernel projection. The processor was a 3.2 GHz Intel® Pentium 4 CPU with 3 GB of random access memory (RAM). Run time was determined using the cputime command.

For each algorithm, computation time increases with increasing training data size (n), as expected. AKFA requires the computation of a Gram matrix whose size increases as the data size increases. The results from the table clearly indicate that AKFA is faster than KPCA. We also noticed that the decrease in computation time for AKFA compared to KPCA was relatively small, implying that the use of AKFA on a smaller training dataset does not yield much advantage over KPCA. However, as the data size increases, the computational gain for AKFA is much larger than that of KPCA as shown in Fig. 2.1. PC-KFA shows more computational time because the composite data-dependent kernels needs calculations of a Gram matrix and optimization of coefficient parameters.

Figure 2.1 illustrates the increase in the computational time of both KPCA as well as AKFA corresponding to increased data. Using Table 2.17, we listed the sizes of all the datasets in the ascending order from lymphoma (77) to colon cancer dataset 4 (3021) on the x-axis versus the respective computational times on the y-axis. The dotted curve indicates the computational time for the KPCA, whereas the dashed curve increases the computational time for AKFA for all eight datasets arranged in ascending order of their sizes. This curve clearly shows that as the size of the data increases, the computational gain of AKFA is greater. This indicates that AKFA is a powerful algorithm and it approaches the performance of KPCA by allowing for significant computational savings.

2.5 CONCLUSION

This study describes first Accelerated Kernel Feature Analysis (AKFA), a faster and more efficient feature extraction algorithm derived from the Sparse Kernel Feature Analysis (SKFA). The time complexity of AKFA is $O(ln^2)$, which has been shown

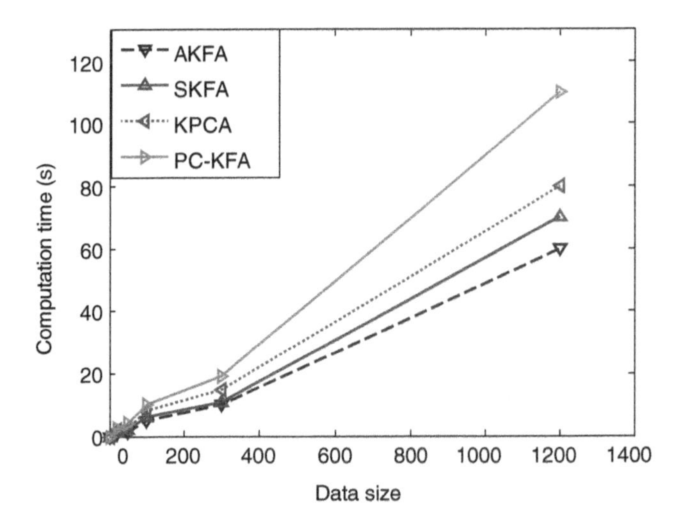

Figure 2.1 The computation time comparison between KPCA, SKFA, AKFA, and PC-KFA as the data size increases.

to be more efficient than the $O(l^2n^2)$ time complexity of SKFA and the complexity $O(n^3)$ of a more systematic PCA (KPCA). We proposed that PC-KFA for AKFA and KPCA. By introducing a principal component metric, the new criteria performed well in choosing the best kernel function adapted to the dataset, as well as extending this process of best kernel selection into additional kernel functions by calculating linear Composite Kernel space. We conducted comprehensive experiments using eight cancer datasets for evaluating the reconstruction error, classification accuracy using a k-nearest neighbor classifier, and computational time. The PC-KFA with KPCA and AKFA had a lower reconstruction error compared to single kernel method, thus demonstrating that the features extracted by the composite kernel method are practically useful to represent the datasets. Composite kernel approach with KPCA and AKFA has the potential to yield high detection performance of polyps resulting in the accurate classification of cancers, compared to the single kernel method. The computation time was also evaluated across the variable data size and this has shown the comparative advantage of composite kernel AKFA.

REFERENCES

1. V. Atluri, S. Jajodia, and E. Bertino, "Transaction processing in multilevel secure databases with kernelized architecture: Challenges and solutions," *IEEE Trans. Knowl. Data Eng.*, vol. 9, no. 5, pp. 697–708, 1997.
2. H. Cevikalp, M. Neamtu, and A. Barkana, "The kernel common vector method: A novel nonlinear subspace classifier for pattern recognition," *IEEE Trans. Syst. Man Cyb. B*, vol. 37, no. 4, pp. 937–951, 2007.
3. G. Horvath and T. Szabo, "Kernel CMAC with improved capability," *IEEE Trans. Syst. Man Cyb. B*, vol. 37, no. 1, pp. 124–138, 2007.

4. C. Heinz and B. Seeger, "Cluster kernels: Resource-aware kernel density estimators over streaming data," *IEEE Trans. Knowl. Data Eng.*, vol. 20, no. 7, pp. 880–893, 2008.
5. B. Schölkopf and A. J. Smola, Learning with kernels. Cambridge, MA: MIT Press, pp. 211–214, 2002.
6. V. N. Vapnik, The nature of statistical learning theory, 2nd ed. New York: Springer, 2000.
7. M. E. Tipping, "Spare kernel principal component analysis," in *Proc. Neural Information Processing Systems (NIPS'00)*, Denver, Colorado, Nov. 28–30, pp. 633–639, 2000.
8. V. France and V. Hlaváč, "Greedy algorithm for a training set reduction in the kernel methods," in *Proc. Computer Analysis of Image and Patterns* (*CAIP 2003*), Groningen, Netherlands, Aug. 25–27, vol. 2756, pp. 426–433, 2003, Lecture Notes in Computer Science.
9. W. Zheng, C. Zou, and L. Zao, "An improved algorithm for kernel principal component analysis," *Neural Process. Lett.*, vol. 22, no. 1, pp. 49–56, 2005.
10. O. L. Mangasarian, A.J. Smola, and B. Schölkopf, "Sparse kernel feature analysis," Tech. Rep. 99–04, University of Wisconsin, 1999.
11. S. Amari and S. Wu, "Improving support vector machine classifiers by modifying kernel functions," *Neural Netw.*, vol. 12, no. 6, pp. 783–789, 1999.
12. Y. Tan and J. Wang, "A support vector machine with a hybrid kernel and minimal Vapnik-Chervonenkis dimension," *IEEE Trans. Knowl. Data Eng.*, vol. 16, no. 4, pp. 385–395, 2004.
13. A. Rakotomamonjy, F. R. Bach, S. Canu, and Y. Grandvalet, "SimpleMKL," *J. Mach. Learn. Res.*, vol. 9, pp. 2491–2521, 2008.
14. R. O. Duda, P. E. Hart, and D.G. Stork, Pattern classification, 2nd ed. Hoboken, NJ: John Wiley & Sons Inc., 2001.
15. J.-B. Pothin and C. Richard, "A greedy algorithm for optimizing the kernel alignment and the performance of kernel machines," in *14th European Signal Processing Conference* (*EUSIPCO 2006*), Florence, Italy, 4–8 September, 2006.
16. J. Kandola, J. Shawe-Taylor, and N. Cristianini, "Optimizing kernel alignment over combinations of kernels," Tech. Rep, NeuroCOLT, 2002.
17. H. Fröhlich, O. Chappelle, and B. Scholkopf, "Feature selection for support vector machines by means of genetic algorithm," in *Proc. 15th. IEEE Int. Conf. Tools with Artificial Intelligence*, Sacramento, California, Nov. 3–5, pp. 142–148, 2003.
18. S. Mika, G. Ratsch, and J. Weston, "Fisher discriminant analysis with kernels," in *Neural Networks for Signal Processing Workshop*, Madison, WI, pp. 41–48, 1999.
19. R. Courant and D. Hilbert, Methods of mathematical physics, Hoboken, NJ: Wiley-VCH, vol. 1, pp. 138–140, 1966.
20. J. Ye, S. Ji, and J. Chen, "Multi-class discriminant kernel learning via convex programming," *J. Mach. Learn. Res.*, vol. 9, pp. 719–758, 1999.
21. B. Souza and A. de Carvalho, "Gene selection based on multi-class support vector machines and genetic algorithms," *Mol. Res.*, vol. 4, no. 3, pp. 599–607, 2005.
22. C. Ding and I. Dubchak, "Multi-class protein fold recognition using support vector machines and neural networks," *Bioinformatics*, vol. 17, no. 4. pp. 349–358, 2001.
23. T. Damoulas and M.A. Girolami, "Probabilistic multi-class multi-kernel learning: On protein fold recognition and remote homology detection," *Bioinformatics*, vol. 24, no. 10, pp. 1264–1270, 2008. doi:10.1093/bioinformatics/btn112.

24. P. Pavlidis, J. Weston, J.S. Cai, et al., "Learning gene functional classifications from multiple data types," *J. Comput. Biol.*, vol. 9, no. 2, pp. 401–411, 2002.
25. O. Chapelle, V. Vapnik, O. Bousquet, et al., "Choosing multiple parameters for support vector machines," *Mach. Learn.*, vol. 46, no. 1–3, pp. 131–159, 2002.
26. C. S. Ong, A. J. Smola, and R. C. Williamson, "Learning the kernel with hyperkernels," *J. Mach. Learn. Res.*, vol. 6, pp. 1043–1071, 2005.
27. G. R. G. Lanckriet, N. Cristianini, P. Bartlett, L.E. Ghaoui, and M.I. Jordan, "Learning the kernel matrix with semidefinite programming," *J. Mach. Learn. Res.*, vol. 5, pp. 27–72, 2004.
28. C. Park and S. B. Cho, "Genetic search for optimal ensemble of feature-classifier pairs in DNA gene expression profiles," in *Proc. Int. Joint Conf. Neural Networks*, Portland, Oregon, July 20–24, vol. 3, pp. 1702–1707, 2003.
29. X. Jiang, R.R. Snapp, Y. Motai, and X. Zhu, "Accelerated kernel feature analysis," in *Proc. IEEE Computer Society Conf. Computer Vision and Pattern Recognition*, New York, New York, June 17–22, pp. 109–116, 2006.
30. K. Chaudhuri, C. Monteleoni, and A.D. Sarwate, "Differentially private empirical risk minimization," *J. Mach. Learn. Res.*, vol. 12, pp. 1069–1109, 2011.
31. K. Duan, S. S. Keerthi, and A. N. Poo, "Evaluation of simple performance measures for tuning SVM hyperparameters," *Neurocomputing*, vol. 51, pp. 41–59, 2003.
32. V. Vapnik and O. Chapelle, "Bounds on error expectation for support vector machines," *Neural Comput.*, vol. 12, no. 9, pp. 2013–2036, 2000.
33. O. Chapelle, V. Vapnik, O. Bousquet, and S. Mukherjee, "Choosing multiple parameters for support vector machines," *Mach. Learn.*, vol. 46, pp. 131–159, 2002
34. K.M. Chung, W.C. Kao, C.-L. Sun, L.-L. Wang, and C.J. Lin,"Radius margin bounds for support vector machines with the RBF kernel," *Neural Comput.*, vol. 15, no. 11, pp. 2643–2681, 2003.
35. J. Yang, Z. Jin, J.Y. Yang, D. Zhang, and A.F. Frangi, "Essence of kernel Fisher discriminant: KPCA plus LDA," *Pattern Recogn.*, vol. 37, pp. 2097–2100, 2004.
36. H. Xiong, Y. Zhang, and X.-W. Chen, "Data-dependent kernel machines for microarray data classification," *IEEE/ACM Trans. Comput. Biol. Bioform.*, vol. 4, no. 4, pp. 583–595, 2007.
37. H. Xiong, M.N.S. Swamy, and M.O. Ahmad, "Optimizing the data-dependent kernel in the empirical feature space," *IEEE Trans. Neural Netw.*, vol. 16, pp. 460–474, 2005.
38. X. W. Chen, "Gene selection for cancer classification using bootstrapped genetic algorithms and support vector machines," in *Proc. IEEE Int. Conf. Computational Systems, Bioinformatics*, Palo Alto, California, Aug. 11–14, pp. 504–505, 2003.
39. H. Yoshida and J. Näppi, "Three-dimensional computer-aided diagnosis scheme for detection of colonic polyps," *IEEE Trans. Med. Imaging*, vol. 20, pp. 1261–1274, 2001.
40. N. Cristianini, J. Kandola, A. Elisseeff, and J. Shawe-Taylor, "On kernel target alignment," in *Proc. Neural Information Processing Systems (NIPS'01)*, Vancouver, British Columbia, Canada, Dec. 3–6, pp. 367–373, 2001.
41. F.A. Sadjadi, "Polarimetric radar target classification using support vector machines," *Opt. Eng.*, vol. 47, no. 4, pp. 046201, 2008.
42. M.A. Shipp, K.N. Ross, P. Tamayo, A.P. Weng, J.L. Kutok, R.C.T. Aguiar, M. Gaasenbeek, M. Angelo, M. Reich, G.S. Pinkus, T.S. Ray, M.A. Koval, K.W. Last, A. Norton, T.A. Lister, J. Mesirov, D.S. Neuberg, E.S. Lander, J.C. Aster, and T.R. Golub, "Diffuse

large B-Cell lymphoma outcome prediction by gene expression profiling and supervised machine learning," *Nat. Med.*, vol. 8, pp. 68–74, 2002.

43. S. Dudoit, J. Fridlyand, and T.P. Speed, "Comparison of discrimination method for the classification of tumor using gene expression data," *J. Am. Stat. Assoc.*, vol. 97, pp. 77–87, 2002.
44. C. Cortes, M. Mohri, and A. Rostamizadeh, "L2 regularization for learning kernels," in *Proceedings of the 25th Conference in Uncertainty in Artificial Intelligence*, Montreal, Canada, June 18–21, 2009.
45. M. Varma and B. R. Babu. More generality in efficient multiple kernel learning, in *Proceedings of the International Conference on Machine Learning*, Montreal, Canada, June 2009.
46. D.J. Newman, S. Hettich, C.L. Blake, and C.J. Merz. UCI repository of machine learning databases, 1998. http://www.ics.uci.edu/~mlearn/MLRepository.html. Accessed 20 November 2014.
47. S. Winawer, R. Fletcher, D. Rex, J. Bond, R. Burt, J. Ferrucci, T. Ganiats, T. Levin, S. Woolf, D. Johnson, L. Kirk, S. Litin, and C. Simmang, "Colorectal cancer screening and surveillance: Clinical guidelines and rationale – Update based on new evidence," *Gastroenterology*, vol. 124, pp. 544–560, 2003.
48. K. D. Bodily, J. G. Fletcher, T. Engelby, M. Percival, J. A. Christensen, B. Young, A. J. Krych, D. C. Vander Kooi, D. Rodysill, J. L. Fidler, and C. D. Johnson, "Nonradiologists as second readers for intraluminal findings at CT colonography," *Acad. Radiol.*, vol. 12, pp. 67–73, 2005.
49. J. G. Fletcher, F. Booya, C. D. Johnson, and D. Ahlquist, "CT colonography: Unraveling the twists and turns," *Curr. Opin. Gastroenterol.*, vol. 21, pp. 90–98, 2005.
50. H. Yoshida and A. H. Dachman, "CAD techniques, challenges, and controversies in computed tomographic colonography," *Abdom. Imaging*, vol. 30, pp. 26–41, 2005.
51. R. M. Summers, C. F. Beaulieu, L. M. Pusanik, J. D. Malley, R. B. Jeffrey, Jr., D. I. Glazer, and S. Napel, "Automated polyp detector for CT colonography: Feasibility study," *Radiology*, vol. 216, pp. 284–290, 2000.
52. R. M. Summers, M. Franaszek, M. T. Miller, P. J. Pickhardt, J. R. Choi, and W. R. Schindler, "Computer-aided detection of polyps on oral contrast-enhanced CT colonography," *AJR Am. J. Roentgenol.*, vol. 184, pp. 105–108, Jan 2005.
53. G. Kiss, J. Van Cleynenbreugel, M. Thomeer, P. Suetens, and G. Marchal, "Computer-aided diagnosis in virtual colonography via combination of surface normal and sphere fitting methods," *Eur. Radiol.*, vol. 12, pp. 77–81, 2002.
54. D. S. Paik, C. F. Beaulieu, G. D. Rubin, B. Acar, R. B. Jeffrey, Jr., J. Yee, J. Dey, and S. Napel, "Surface normal overlap: A computer-aided detection algorithm with application to colonic polyps and lung nodules in helical CT," *IEEE Trans. Med. Imaging*, vol. 23, pp. 661–675, 2004.
55. A. K. Jerebko, R. M. Summers, J. D. Malley, M. Franaszek, and C. D. Johnson, "Computer-assisted detection of colonic polyps with CT colonography using neural networks and binary classification trees," *Med. Phys.*, vol. 30, pp. 52–60, 2003.
56. J. Näppi, H. Frimmel, A. H. Dachman, and H. Yoshida, "A new high-performance CAD scheme for the detection of polyps in CT colonography," in *Medical Imaging 2004: Image Processing*, San Diego, CA, February 14, pp. 839–848, 2004.

57. A. K. Jerebko, J. D. Malley, M. Franaszek, and R. M. Summers, "Multiple neural network classification scheme for detection of colonic polyps in CT colonography data sets," *Acad. Radiol.*, vol. 10, pp. 154–160, 2003.
58. A. K. Jerebko, J. D. Malley, M. Franaszek, and R. M. Summers, "Support vector machines committee classification method for computer-aided polyp detection in CT colonography," *Acad. Radiol.*, vol. 12, pp. 479–486, 2005.
59. J. Franc and V. Hlavac, "Statistical Pattern Recognition Toolbox for MATLAB, 2004," http://cmp.felk.cvut.cz/cmp/software/stprtool/. Accessed 20 November 2014.
60. "Partners Research Computing," [On-line], 2006, http://www.partners.org/rescomputing/. Accessed 20 November 2014.
61. A. J. Smola and B. Schölkopf, "Sparse greedy matrix approximation for machine learning," in *Proc. 17th Int. Conf. Machine Learning*, Stanford, California, June 29–July 2, 2000.
62. K. Fukunaga and L. Hostetler, "Optimization of k-nearest neighbor density estimates," *IEEE Trans. Inform. Theor.*, vol. 19, no. 3, pp. 320–326, 1973.
63. J. H. Friedman, Flexible metric nearest neighbor classification. Technical report, Stanford, CA, USA: Department of Statistics, Stanford University, November 1994.
64. T. Hastie and R. Tibshirani, "Discriminant adaptive nearest neighbor classification," *IEEE Trans. Pattern Anal. Mach. Intell.*, vol. 18, no. 6, pp. 607–616, 1996.
65. D. G. Lowe, "Similarity metric learning for a variable-kernel classifier," *Neural Comput.*, vol. 7, no. 1, pp. 72–85, 1995.
66. J. Peng, D.R. Heisterkamp, and H.K. Dai, "Adaptive kernel metric nearest neighbor classification," in *Proc. Sixteenth Int. Conf. Pattern Recognition*, Québec City, Québec, Canada, vol. 3, pp. 33–36, 11–15 August, 2002.
67. Q. B. Gao and Z. Z. Wang, "Center-based nearest neighbor classifier," *Pattern Recogn.*, vol. 40, pp. 346–349, 2007.
68. S. Li and J. Lu, "Face recognition using the nearest feature line method," *IEEE Trans. Neural Netw.*, vol. 10, no. 2, pp. 439–443, 1999.
69. P. Vincent, Y. Bengio, "K-local hyperplane and convex distance nearest neighbor algorithms," in *Advances in Neural Information Processing Systems* (*NIPS*), vol. 14. Cambridge, MA: MIT Press, pp. 985–992, 2002.
70. W. Zheng, L. Zhao, and C. Zou, "Locally nearest neighbor classifiers for pattern classification," *Pattern Recogn.*, vol. 37, pp. 1307–1309, 2004.
71. G. C. Cawley, MATLAB Support Vector Machine Toolbox. Anglia: School of Information Systems, Univ. of East, 2000, http://theoval.cmp.uea.ac.uk/svm/toolbox/, Norwich, UK.
72. Y. Raviv and N. Intrator, "Bootstrapping with noise: An efficient regularization technique," *Connect. Sci.*, vol. 8, pp. 355–372, 1996.
73. H.A. Buchholdt, Structural dynamics for engineers. London, England: Thomas Telford, 1997, ISBN 0727725599.
74. T. Briggs and T. Oates, "Discovering domain specific composite kernels" in *Proc. 20th National Conf. Artificial Intelligence and 17th Annual Conf. Innovative Applications Artificial intelligence*, Pittsburgh, Pennsylvania, July 9–13, pp. 732–739, 2005.
75. C. Sotiriou, P. Wirapati, S. Loi, A. Harris, et al., "Gene expression profiling in breast cancer: Understanding the molecular basis of histologic grade to improve prognosis" *J. Natl. Cancer Inst.*, vol. 98, no. 4, pp. 262–272, 2006. PMID: 16478745.

76. U.R. Chandran, C. Ma, R. Dhir, M. Bisceglia, et al., "Gene expression profiles of prostate cancer reveal involvement of multiple molecular pathways in the metastatic process," *BMC Cancer*, vol. 7, pp. 7–64, 2007. PMID: 17430594.
77. Y.P. Yu, D. Landsittel, L. Jing, J. Nelson, et al., "Gene expression alterations in prostate cancer predicting tumor aggression and preceding development of malignancy," *J. Clin. Oncol.*, vol. 22, no. 14, pp. 2790–2799, 2004; PMID: 15254046.

3

GROUP KERNEL FEATURE ANALYSIS[1]

3.1 INTRODUCTION

The major concern in traditional medical screening is that it is a limited evaluation by a physician who may diagnose patients on the basis of his/her knowledge. In the case of colon cancer, although many patients received conventional endoscope screenings, it was estimated that it is the third most common cancer in the United States, causing over 136,000 deaths every year [1–3]. For screening, computer tomographic colonography (CTC) is emerging as an attractive alternative to more invasive colonoscopy because it can find benign growths of polyps in their early stages, so that they can be removed before cancer has had a chance to develop [4]. Improvements that reduce the diagnosis error would go a long way toward making CTC a more acceptable technique by using computer tomography (CT) scans of the colon [1]. To be a clinically practical means of screening colon cancers, CTC must be able to interpret a large number of images in a time-effective manner, and it must facilitate the detection of polyps—precursors of colorectal cancers—with high accuracy. Currently, however, interpretation of an entire CTC examination is handled by a limited number of specialists at individual hospitals, and the reader performance for polyp detection varies substantially [1, 3]. To overcome these difficulties while providing accurate detection of polyps, computer-aided detection (CAD) schemes are investigated that semiautomatically detect suspicious lesions in CTC images [3].

The state of the art of CAD is emerging as CTC gains popularity for cancer diagnosis. Numerical schemes of image analysis have been developed for individual

[1]This chapter is a revised version of the author's paper in ACM Transactions on Intelligent Systems and Technology, approved by ACM.

Data-Variant Kernel Analysis, First Edition. Yuichi Motai.

institutions (i.e., hospitals), where resources and training requirements determine the number of training instances [4–9]. Thus, if more training data are collected after the initial tumor model is computed, retraining of the model becomes imperative in order to incorporate the data from other institutions and to preserve or improve the classification accuracy [5, 6, 10, 11]. Hereafter, we propose a new framework, called "distributed colonography," where the colonography database at each institution may be shared and/or uploaded to a common server. The CAD system at each institution can be enhanced by incorporating new data from other institutions by using the distributed learning model proposed in this study.

The concept of distributed colonography using networked distributed databases has been discussed in many classification applications but not yet in the context of CAD in CTC [12, 13]. These existing studies showed that the overall classification performance for larger multiple databases was improved in practical settings [14, 15]. Rather than applying traditional techniques of classification to the very limited number of patients, medical data from several institutions will be expected. The utilization of the proposed distributed colonography framework shown in Fig. 3.1 requires a comprehensive study to determine whether the overall performance is improved by using multiple databases. The presented work is a first attempt at such a study. The benefit for clinical practice is that different types of cancer datasets will be available, which may not exist at one specific institution. The number of true positive cases is usually small, because cancer screenings are very general, and most screenings are of healthy, noncancer people. The CAD algorithm and clinicians both can observe all of the potential cases in the proposed distributed platform [16, 17].

The primary focus of this study is to find effective ways to associate multiple databases to represent the statistical data characteristics. Few existing classification techniques using distributed databases successfully handle big data structures. The new classification method is expected to be capable of learning multiple large databases, specifically tailored for the large amount of CTC data. Thus, we propose

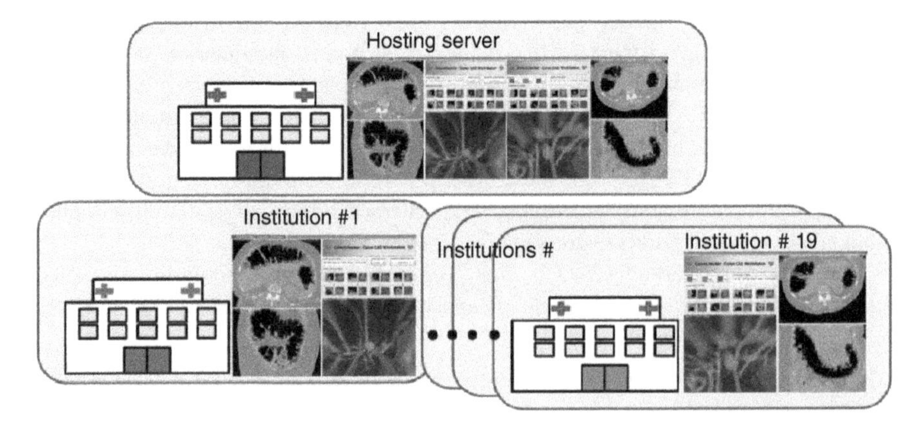

Figure 3.1 Distributed colonography with distributed image databases for colon cancer diagnosis. The hosting server collects and analyzes databases from different institutions and groups them into assembled databases.

composite kernel feature analysis with multiple databases for this CTC problem due to the effective compression of big data. The improvement in performance that could be obtained by employing the proposed approach requires very small additional investment in terms of resources, infrastructure, and operational costs.

We propose group kernel feature analysis (GKFA) for distributed databases for the group learning method. GKFA can efficiently differentiate polyps from false positives and thus can be expected to improve detection performance. The key idea behind the proposed GKFA method is to allow the feature space to be updated as the training proceeds with more data being fed from other institutions into the algorithm. The feature space can be reconstructed by GKFA by grouping multiple databases. The feature space is augmented with new features extracted from the new data, with a possible expansion to the feature space if necessary. We present the first comparative study of these methods and show that the proposed GKFA outperforms existing nonlinear dimensionality reduction methods when different databases for CTC become necessary.

The contribution of this study is that the proposed GKFA method works in distributed CTC databases. These databases acquired over a long period of time can sometimes be highly diverse, and each database is unique in nature; therefore, obtaining a clear distinction among multiple databases is a very challenging task. There is a chance of misinterpretation of the database to be either homogeneous or heterogeneous in nature while training the new incoming databases from many institutions. The method was tested using real CTC data to show that the proposed GKFA improves the CAD performance while achieving a feature space that is comparably similar to that obtained by a separate learning method at each institution.

Each database of each hospital consists of many data sources of different patients from various demographic distributions, such as ages, genders, races, and ethnicities. We have analyzed the heterogeneity of these data variations in Section 3.3.2, determining the degree of heterogeneity as homogeneous, small heterogeneous, or large heterogeneous. To find effective ways to handle big datasets from multiple databases, the proposed GKFA uses heterogeneous/homogeneous criteria to evaluate the statistical data characteristics to combine the databases.

The rest of this work is organized as follows. Section 3.2 provides an introduction to kernel methods and a brief review of the existing kernel-based feature extraction methods kernel principal component analysis (KPCA). In Section 3.3, we discuss homogeneous and heterogeneous grouping of subsets of databases for the problem of huge size of incoming data. Section 3.4 describes the proposed GKFA for the polyp candidates from multiple databases, Section 3.5 evaluates the experimental results, while conclusions are drawn in Section 3.6.

3.2 KERNEL PRINCIPAL COMPONENT ANALYSIS (KPCA)

There are many existing feature selection methods for classifiers such as stepwise feature selection [18] and manifold-based methods [19] in the generic field of pattern recognition. It is not certain if these existing methods are effective when used in the proposed distributed CTC framework. The effective application of group learning using multiple databases to nonlinear spaces is undertaken by using kernel-based methods [12, 13, 20–22]. Kernel-based feature extraction methods tend to perform better than non-kernel-based methods because the actual databases have very nonlinear characteristics. Another issue under consideration is the approach used to handle the larger sized databases obtained by combining multiple databases. In [21], Zheng proposed that the input data be divided into a few groups of similar size and then KPCA is applied to each group. A set of eigenvectors was obtained for each group, and the final set of features was obtained by applying KPCA to a subset of these eigenvectors. The application of KPCA is referred to as the most promising method of compressing all the databases and extracting the salient features (principal components) [23–27]. KPCA has already shown computational effectiveness in many image processing applications and pattern classification systems [28–33].

For efficient feature analysis, extraction of the salient features of polyps is essential because of the size and the 3D nature of the polyp databases [10]. Moreover, the distribution of the image features of polyps is nonlinear [10]. The problem is how to select a nonlinear, positive-definite kernel $K: R^d \times R^d \rightarrow R$ of an integral operator in the d-dimensional space. The kernel K, which is a Hermitian and positive semidefinite matrix, calculates the inner product between two finite sequences of inputs $\{x_i : i \in n\}$ and $\{x_j : j \in n\}$, defined as $K := (K(x_i, x_j)) = (\Phi\ (x_i).\ \Phi\ (x_j) : i,j \in n)$. Here, x is a gray-level CT image, n is the number of image databases, and $\Phi : \ R^d \rightarrow H$ denotes a nonlinear embedding (induced by K) into a possibly infinite dimensional Hilbert space H. Some of the commonly used kernels are the linear kernel, the polynomial kernel, the Gaussian radial basis function (RBF) kernel, the Laplace RBF kernel, the Sigmoid kernel, and the analysis of variance (ANOVA) RB kernel [34].

Kernel selection is heavily dependent on the data specifics. For instance, the linear kernel is important in large sparse data vectors and it implements the simplest of all kernels, whereas the Gaussian and Laplace RBFs are general purpose kernels used when prior knowledge about data is not available. The Gaussian kernel avoids the sparse distribution, which is obtained when a high-degree polynomial kernel is used. The polynomial kernel is widely used in image processing, while the ANOVA RB is usually adopted for regression tasks. A more thorough discussion of kernels can be found in [34–38]. Our GKFA for CTC images is a dynamic extension of KPCA as follows: [28–31, 39, 40].

KPCA uses a Mercer kernel [6] to perform a linear principal component analysis of the transformed image. Without loss of generality, we assume that the image of the data has been centered so that its scatter matrix in S is given by $S = \sum_{i=1}^{n} \Phi\ (x_i)\ (x_i)\ \Phi\ (x_i)^T$. The eigenvalues λ_j and eigenvectors e_j are obtained by solving the following equation, $\lambda_j e_j = Se_j = \sum_{i=1}^{n} \Phi\ (x_i)\ \Phi\ (x_i)^T e_j = \sum_{i=1}^{n} \langle e_j, \Phi\ (x_i) \rangle \Phi\ (x_i)$. If K is an $n \times n$ Gram matrix, with the element $k_{ij} = \ \langle \Phi(x_i), \Phi(x_j) \rangle$ and $a_j = [a_{j1}, a_{j2}, \ldots, a_{jn}]$ are the eigenvectors associated with eigenvalues λ_j, then the dual eigenvalue problem equivalent to the problem can be expressed as follows: $\lambda_j a_j = Ka_{j.}$

KPCA can now be presented as follows:

1. Compute the Gram matrix that contains the inner products between pairs of image vectors.
2. Solve $\lambda_j a_j = K a_j$ to obtain the coefficient vectors a_j for $j = 1, 2, \ldots, n$.
3. The projection of a test point $x \in R^d$ along the jth eigenvector is $\langle e_j, \Phi(x) \rangle = \sum_{i=1}^{n} a_{ji} \langle \Phi(x_i), \Phi(x) \rangle = \sum_{i=1}^{n} a_{ji} k(x, x_i)$.

The above equation implicitly contains an eigenvalue problem of rank n, so the computational complexity of KPCA is $O(n^3)$. The total computational complexity is given by $O(ln^2)$ where l stands for the number of features to be extracted and n stands for the rank of the Gram matrix K [39, 40]. Once the Gram matrix is computed, we can apply these algorithms to our database to obtain a higher dimensional feature space. This idea is discussed in the following sections.

3.3 KERNEL FEATURE ANALYSIS (KFA) FOR DISTRIBUTED DATABASES

Nonshareable data have not yet been addressed in any clinical applications of colonography CAD [4, 8, 9, 41]. In the current environment, the problem is that each platform is independently operated in a closed manner—that is, none of the CAD platforms contributes to other CAD platforms. To avoid this limitation, the proposed distributed databases for colonography attempts to make larger data-driven CAD more prominent with the data aggregation. To handle the data aggregation by synthesizing each platform, instead of handling data independently, we herein propose a machine learning technique, called GKFA that adjusts the classification criteria by extending KPCA for distributed colonography databases. We want to validate the performance of GKFA when applied to a distributed database, specifically CTC.

As shown in Fig. 3.2, we introduce the concept of training the algorithm by analyzing the data received from other databases. Step 1 in Fig. 3.2 illustrates the decomposition of each database through kernel feature analysis (KFA). Each database consists of several datasets. For example, database #1 is decomposed into four datasets. We will describe Step 1 in Section 3.3.1 and Step 2 in Section 3.3.2 to reconstruct each database using KFA. Specifically, Section 3.3.1 describes how to extract the data-dependent kernels for each database using KFA. In Section 3.3.2, we propose that each database is classified as either homogeneous or heterogeneous by the proposed criteria, so that each database can be decomposed into a heterogeneous dataset. We describe details on Step 3 separately in Section 3.4 as GKFA.

3.3.1 *Extract Data-Dependent Kernels Using KFA*

We exploit the idea of the data-dependent kernel to select the most appropriate kernels for a given database. Let $\{x_i, x_j\} (i, j = 1, 2, \ldots, n)$ be n training samples

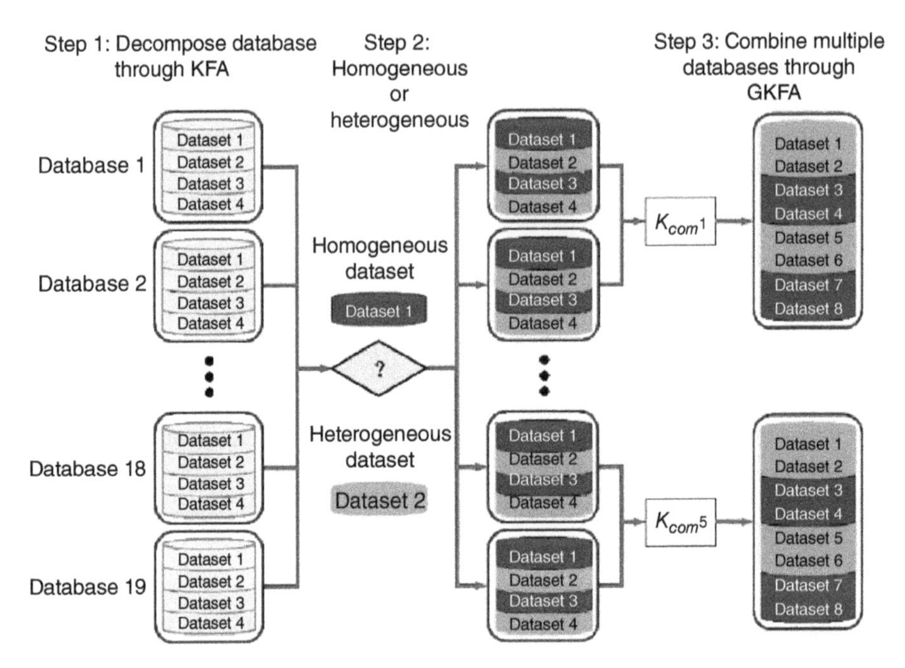

Figure 3.2 The concept of group kernel feature analysis. The proposed criteria are to determine the nature of database by (i) decomposition, (ii) classification by heterogeneity, and (iii) combination.

of the given d-dimensional data, and $x_j = \{+1, -1\}$ represents the class labels of the samples, that is, the data is labeled true positive representing $x_j = \{+1\}$. A data-dependent kernel is adopted using a composite form, as follows: the kernel k_l for $l \in \{1, 2, 3, 4\}$ is formulated as $k_l(x_i, x_j) = q_l(x_i)q_l(x_j)p_l(x_i, x_j)$, where, $x \in R^d$, $p_l(x_i, x_j)$ is one kernel among four chosen kernels. Here, $q_l(\cdot)$ is the factor function, $q_l(x_i) = a_{l0} + \sum_{m=1}^{n} a_{lm}k_l(x_i, a_{lm})$, where $k_l(x_i, x_j)$, and α_{lm} are the combination coefficients. In matrix form, we can write $q_l = K_l \cdot a$, where $q_l = \{q_l(x_1), q_l(x_2), \ldots, q_l(x_n)\}^T$ and K_l is an $n \times (n+1)$ matrix defined as

$$K_l = \begin{pmatrix} 1 & k_l\left(x_1, a_{l1}\right) & \cdots & k_l(x_1, a_{ln}) \\ 1 & k_l(x_2, a_{l1}) & \cdots & k_l(x_2, a_{ln}) \\ \vdots & \vdots & \cdots & \vdots \\ 1 & k_l(x_n, a_{l1}) & \cdots & k_l(x_n, a_{ln}) \end{pmatrix} \tag{3.1}$$

Let the kernel matrices corresponding to kernels k_l, and p_l be K_l, and P_l. Therefore, we can denote the data-dependent kernel matrix K_l as $K_l = [q_l(x_i)q_l(x_j)p_l(x_i, x_j)]_{n\times(n+1)}$. Defining $Q_l = \{1, q_l(x_1), q_l(x_2), \ldots, q_l(x_n)\}^T$, we obtain $K_l = Q_lP_lQ_l^T$. We decompose each database by maximizing the Fisher scalar for our kernel optimization. The Fisher scalar is used to measure the class separability J of the training data in the mapped feature space. It is formulated as $J = \mathrm{tr}(\Sigma_l S_{bl}) / \mathrm{tr}\left(\sum_l S_{wl}\right)$, where S_{bl} represents "between-class scatter matrices" and S_{wl} represents "within-class scatter matrices."

Suppose that the training data are clustered, that is, the first n_1 data belong to one class (class label equals −1), and the remaining n_2, data belong to the other class (class label equals +1). Then, the basic kernel matrix P_l. can be partitioned to represent each class shown as

$$P_l = \begin{pmatrix} P^l_{11} & P^l_{12} \\ P^l_{21} & P^l_{22} \end{pmatrix} \tag{3.2}$$

where ${P_{11}}^l, {P_{12}}^l, {P_{21}}^l$, and ${P_{22}}^l$ are the submatrices of P_l in the order of $n_1 \times n_1, n_1 \times n_2, n_2 \times n_1, n_2 \times n_2$, respectively. According to [42], the class separability by the Fisher scalar can be expressed as $J_l(a_l) = a_l^T M_l a_l / a_l^T N_l a_l$, where $M_l = K_l^T B_l K_l$, $N_l = K_l^T W_l K_l$, and using diag(.) to extract a principal diagonal of the matrix

$$W_l = \text{diag}(P^l_{11}, P^l_{22}) - \begin{pmatrix} P^l_{11}/n_1 & 0 \\ 0 & P^l_{22}/n_2 \end{pmatrix} \quad B_l = \begin{pmatrix} P^l_{11}/n_1 & 0 \\ 0 & P^l_{22}/n_2 \end{pmatrix} - \frac{P_l}{n} \tag{3.3}$$

To maximize $J_l(a_l)$, the standard gradient approach is followed. If the matrix N_{0i} is nonsingular, the optimal a_l which maximizes the $J_l(a_l)$ is the eigenvector that corresponds to the maximum eigenvalue of $M_l a_l = \lambda_l N_l a_l$. The criterion to select the best kernel function is to find the kernel that produces the largest eigenvalue.

$$\lambda_{l*} = \arg\max_{\lambda_l}(N_l^{-1} M_l) \tag{3.4}$$

Choosing the eigenvector that corresponds to the maximum eigenvalue can maximize the $J_l(a_l)$ to achieve the optimum solution. Once we determine the eigenvectors, that is, the combination coefficients of all four different kernels, we now proceed to construct q_l and Q_l to find the corresponding Gram matrices of kernel K_l.

The optimization of the data-dependent kernel k_l consists of selecting the optimal combination coefficient vector a_l so that the class separability of the training data in the mapped feature space is maximized. Once we have computed these Gram matrices, we can find the optimum kernels for the given database. To do this, we arrange the eigenvalues (that determined the combination coefficients) for all kernel functions in descending order. The first kernel corresponding to the largest eigenvalue is used in the construction of a composite kernel that is expected to yield the optimum classification accuracy [43]. If we apply one of the four kernels for the entire database, we cannot achieve the desired classification performance [14].

3.3.2 Decomposition of Database Through Data Association via Recursively Updating Kernel Matrices

As shown in Step 2 of Fig. 3.2, in each database, we apply the data association using class separability as a measure to identify whether the data is heterogeneous or homogeneous. Obtaining a clear distinction between heterogeneous and homogeneous data of each database is a very challenging task. The data acquired in clinical trials can sometimes be highly diverse, and there is no concrete method to differentiate between heterogeneous and homogeneous data. We would like to decompose each database so that the decomposed database has homogeneous data characteristics. If the data is homogeneous, then the separability is improved, and conversely, the heterogeneous data degrades the class separability ratio. Let us introduce a variable ξ, which is the ratio of the class separabilities:

$$\xi = \arg\max_{r} \left(\frac{J'_{*}(a'_{r})}{J_{*}(a'_{r})} \right) \tag{3.5}$$

where $J_{*}'(a_l') = a_l'^T M_l' a_l' / a_l'^T N_l' a_l'$ denotes the class separability yielded by the most dominant kernel, which is chosen from the four different kernels for the dataset, and $J_{*}(a_l') = a_l'^T M_l a_l' / a_l'^T N_l a_l'$ is the class separability yielded by the most dominant kernel for the entire database. Because of the properties of class separability, Equation 3.5 can be rewritten as $\xi = \lambda_{*}'/\lambda_{**}$, where $\lambda_{*'}$ correspond to the most dominant eigenvalue of in the maximization of Equation 3.4 under the kernel choice, given all clusters noted as r clusters. The newly computed λ_{*}' is the latest eigenvalue of another r set of "recalculated" clusters. If ξ is less than a threshold value *1*, then the database is heterogeneous; otherwise, it is homogeneous. If the data is homogeneous, we keep the Gram matrix, as it is defined in Section 3.3.1. Conversely, if the data is heterogeneous, we update the Gram matrix depending on the level of heterogeneity of the subdatasets, as described in the following section.

We propose to quantify the data's heterogeneity by introducing a criterion called the "residue factor" by extending Equation 3.5 $\xi = \lambda_{*}'/\lambda_{*}$ into the residue factor "***rf***," defined as

$$\boldsymbol{rf} = (\boldsymbol{a}'_{*} - \overrightarrow{\boldsymbol{a}}_{*}) \cdot \frac{\lambda'_{*}}{\lambda_{*}} \tag{3.6}$$

where we use only the most dominant kernel for determining the residue factor. The class separability of the most dominant kernel for the newly decomposed data is directly dependent on both the maximum combination coefficient a'_{*} (this is the maximum combination coefficient of four different kernels), as well as the maximum eigenvalue λ'_{*}. Let us denote by $\overrightarrow{a}_{*}$ the mean of the combination coefficients of all database, and by a'_{*} the most dominant kernel among the subsets of the newly decomposed database, respectively. Using these values, we determine the type of update for the Gram matrix by evaluating disparities between composite eigenvalues, iteratively.

We have observed that there exists a chance of misinterpretation of the heterogeneous data while updating the newly clustered databases. Specifically, shown in Fig. 3.3, we consider the two cases of the updates for small/large heterogeneous data as follows:

Case 1: Partially Update for Small Heterogeneous Data If the residue factor ***rf*** in Equation 3.6 is less than a threshold value η, then that means the heterogeneous degree between the previous eigenvectors and the new eigenvectors is relatively small. Hence, the dimensions of the Gram matrix have to remain constant. We replace the trivial rows and columns of the dominant kernel Gram matrix with those of the newly decomposed data. The trivial rows/columns are calculated by the

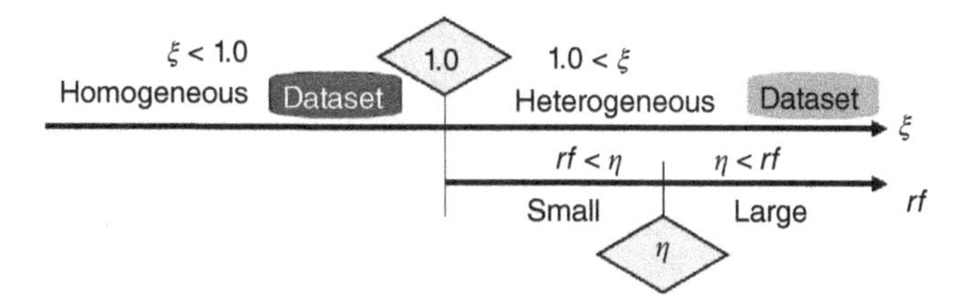

Figure 3.3 Relationship of the criteria to determine homogenous and heterogeneous degree.

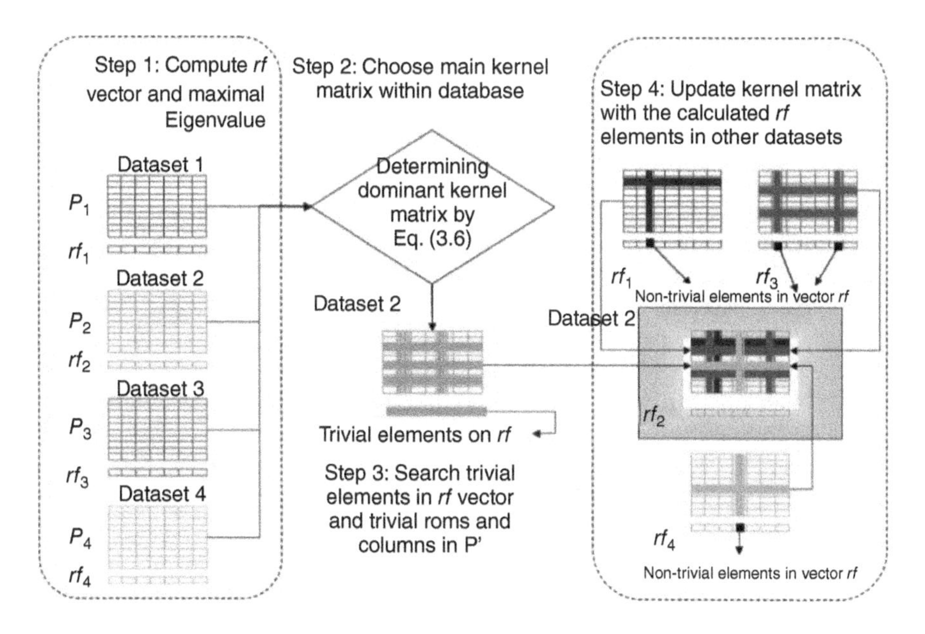

Figure 3.4 The steps to choose the basic kernel matrix P' and the updating process according to the elements rf.

minimum difference vector. As we assigned the decomposed data to one of the existing datasets, we just compare the combination coefficient values of that class with the combination coefficient of new decomposed data to yield the difference vectors that determine the trivial combination vector to be replaced. This process is repeated for all the kernel matrices.

The input matrices P_l' and Q_l' should also be updated by removing rows and columns by applying the four steps shown in Fig. 3.4, as follows.

In Step 1, we compute the individual residue factor $\boldsymbol{rf}$ corresponding to each dataset, decompose one matrix corresponding to one database into kernel matrices for several datasets. In Step 2, we choose the main kernel matrix among the datasets by maximizing ξ. In Step 3, we search trivial elements in the $\boldsymbol{rf}$ vector in the main kernel matrix according to Equation 3.6 that minimizes $\boldsymbol{rf}$. In Step 4, we substitute the corresponding parts in the main kernel matrix with the calculated element of ***rf*** in the other datasets. We compute $Q_l' = \text{diag}(a_l')$. Hence, in the update for small heterogeneous data, the Gram matrix can be given as $K_l' = Q_l' P_l' Q_l'$.

Case 2: Update for Large Heterogeneous Data If the residue factor ***rf*** in Equation 3.6 is greater than a threshold value η, that means the heterogeneous degree between the previous eigenvectors and the new eigenvectors is relatively large. So it is very important for us to retain this highly heterogeneous data for efficient classification. Instead of replacing the trivial rows and columns of the previous data, we simply retain them. Hence, the size of the Gram matrix is increased by the size of the newly decomposed data. We already calculated the new combination coefficient a_l', and input matrix P' and Q' are as same as in Section 3.3.1. Then, the kernel Gram matrix can be newly calculated as

$$K_l^{n'} = Q_l' P_l' Q_l' \tag{3.7}$$

Once we have our kernel Gram matrices, we can now determine the composite kernel that gives us the optimum classification accuracy when the existing database is incorporated with newly decomposed data.

We perform the entire algorithm to see if there is an improvement in the difference of $\vec{a}_*$ and a_*' from the current and previous step. If the heterogeneous degree is still large, then the decomposed data has to be further reduced, and the recursive algorithm described herein is performed again. This entire process is summarized in the flow chart in Fig. 3.5. This process is repeated until the residue factor finds an appropriate size of data that would allow for all the decomposed datasets to be homogeneous. That means the residue factor is expected to converge to zero by recursively updating clusters:

$$\lim_{n\to\infty} (\arg(rf^n)) \approx 0 \tag{3.8}$$

After the training of the Gram matrix is finished to incorporate the heterogeneous/homogeneous features of the newly decomposed data, the KFA algorithm is applied to the kernel Gram matrix. Obtaining a higher dimensional feature space of the huge data with greater classification power depends on how effectively we are updating the Gram matrix. As the size of distributed databases increases, it is often very important to reevaluate and change the criteria established in an enhanced algorithm to correctly train the big data. In the following section, we explore how the reclustered multiple databases must be aligned.

3.4 GROUP KERNEL FEATURE ANALYSIS (GKFA)

There is currently insufficient true cancer data (leading to low sensitivity) for a full colon-cleaning preparation in any single medical database to validate the accuracy and performance of CTC. Owing to the limitation of training cases, few radiologists have clinical access to the full variety of true positive cases encountered in actual

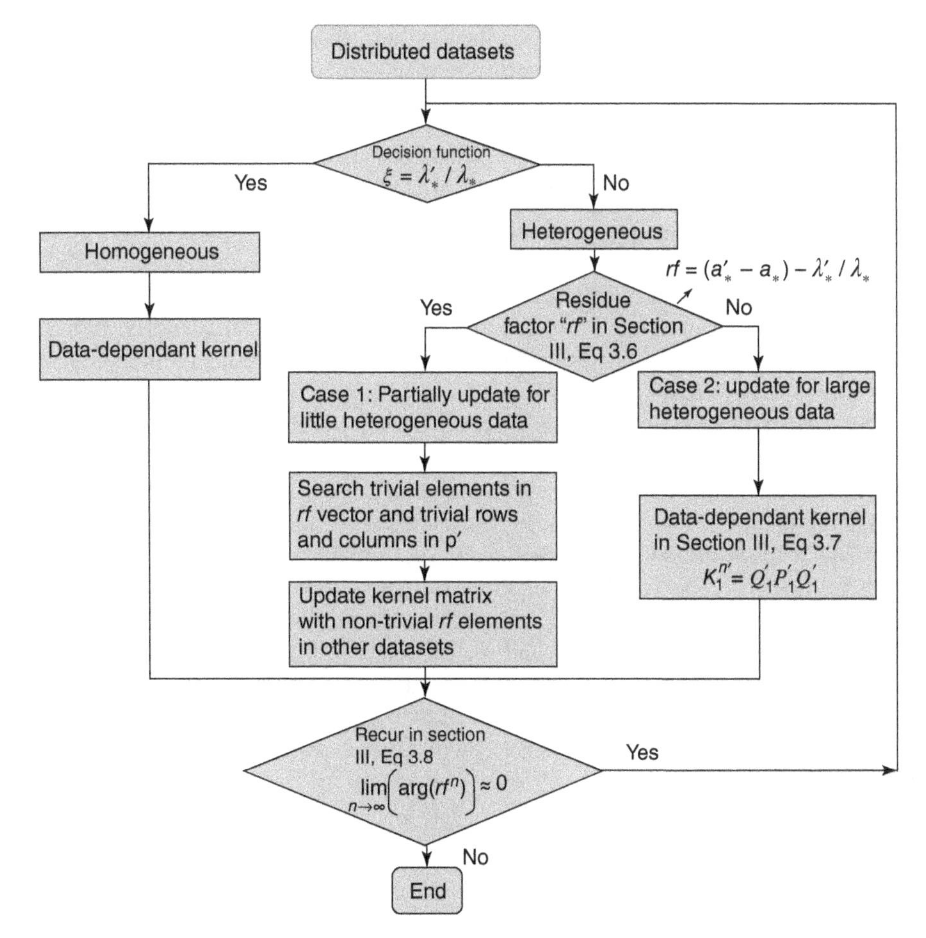

Figure 3.5 The training flow chart of reclustered databases due to the heterogeneous nature.

clinic practices of CTC. Therefore, we propose combining the data from several databases to improve classification by increasing the colon cancer cases and diversities of the patients [5, 6, 10, 11, 17]. There are as of yet few data centers that render data into a form that can be readily reused, shoulder curatorial responsibilities, or build new data management tools and services.

In this section, as shown in Step 3 of Fig. 3.2, we illustrate how to handle "multiple databases" through GKFA.

3.4.1 Composite Kernel: Kernel Combinatory Optimization

In this section, we propose a composite kernel function to define the weighted sum of the set of different optimized kernel functions, which correspond to multiple clustered databases. To obtain the optimum classification accuracy, we define the composite kernel $K^s_{com}(\rho)$, using a composite coefficient ρ, as

$$K^s_{com}(\rho) = \rho_{l_1} Q_{l_1} P_{l_1} Q_{l_1} + \rho_{l_2} Q_{l_2} P_{l_2} Q_{l_2} \tag{3.9}$$

where $K^s_{com}(\rho)$ is extended from the composite kernel with the variable of the combinatory number that we intend to combine two basic kernel functions among four kernels: the linear kernel, the polynomial kernel, the Gaussian RBF kernel, and the Laplace RBF. That means that the number of possible combinations is six, that is, ${}_4C_2 = 6$ cases, so s represents one of six composite kernels. Through this approach, the relative contribution of a single kernel to the composite kernel can be varied over the multiple databases by the value of the composite coefficient ρ. Instead of using K_l as the kernel matrix, we will be using $K^s_{com}(\rho)$. According to [14], this composite kernel matrix $K^s_{com}(\rho)$ satisfies Mercer's condition.

The problem becomes how to determine this composite coefficient $\widehat{\rho}$ ($\widehat{\rho} = [\rho_{l_1}, \rho_{l_2}]$) such that the classification performance is optimized. To this end, we used the concept of "kernel alignment" to determine the best $\widehat{\rho}$ that gives us optimum performance. The alignment measure was proposed by Cristianini and Kandola [44] to compute the adaptability of a kernel to the target data and provide a practical method to optimize the kernel. It is defined as the normalized Frobenius inner product between the kernel matrix and the target label matrix. The empirical alignment between kernel k_1 and kernel k_2 with respect to the training set is given as

$$A(k_1, k_2) = \frac{\langle K_1, K_2 \rangle_F}{\|K_1\|_F \|K_2\|_F} \tag{3.10}$$

where K_1 and K_2 are the kernel matrix for the training set using kernel function k_1 and k_2, $\|K_1\|_F = \sqrt{\langle K_1, K_1 \rangle_F}$, $\|K_2\|_F = \sqrt{\langle K_2, K_2 \rangle_F}$. $\langle K_1, K_2 \rangle_F$ is the Frobenius inner product between K_1 and K_2. If $K_2 = yy^T$, then the empirical alignment between kernel K^s_{com} and target vector y is

$$\begin{aligned} A(K^s_{com}, yy^T) &= \frac{\langle K^s_{com}, yy^T \rangle_F}{\sqrt{\langle K^s_{com}, K^s_{com} \rangle_F \langle yy^T, yy^T \rangle_F}} \\ &= \frac{y^T K^s_{com} y}{n\sqrt{\langle K^s_{com}, K^s_{com} \rangle_F}} \end{aligned} \tag{3.11}$$

If the kernel is well adapted to the target information, separation of the data has a low bound on the generalization error [44]. So, we can optimize the kernel alignment by training data to improve the generalization performance on the testing data. Let us consider the optimal composite kernel corresponding to Equation 3.9 as $K^s_{com}(\widehat{\rho}) = \widehat{\rho}_{l_1} Q_{l_1} P_{l_1} Q_{l_1} + \widehat{\rho}_{l_2} Q_{l_2} P_{l_2} Q_{l_2}$. We can change ρ to maximize the empirical alignment between $K^s_{com}(\hat{\rho})$ and the target vector yy^T. Hence,

$$\rho = \arg\max(A(K^s_{com}, yy^T))$$
$$= \arg\max\left(\frac{\left\langle \sum_{l=1}^{2} \rho_l K_l, yy^T \right\rangle}{n\sqrt{\left\langle \sum_{l_1=1}^{2} \rho_{l_1} K_{l_1} \right\rangle, \left\langle \sum_{l_2=1}^{2} \rho_{l_2} K_{l_2} \right\rangle}}\right)$$
$$= \arg\max\left(\frac{\left(\sum_{l=1}^{2} \rho_l u_l\right)^2}{n^2 \sum_{l_1=1,l_2=1}^{2} \rho_{l_1}\rho_{l_2} v_{l_1 l_2}}\right) \quad (3.12)$$
$$= \arg\max\left(\frac{\rho^T U \rho}{n^2 \rho^T V \rho}\right) \quad (3.13)$$

where $u_l = \sqrt{\langle K_l, yy^T\rangle}, v_{l_1 l_2} = \sqrt{\langle K_{l_1}, K_{l_2}\rangle}, U_{l_1 l_2} = u_{l_1} u_{l_2} V_{l_1 l_2} = v_{l_1} v_{l_2}, \rho = (\sqrt{\rho_{l_1}}, \sqrt{\rho_{l_2}})$. Let the generalized Raleigh coefficient be $J(\rho) = \rho^T U \rho / \rho^T V \rho$. Therefore, we can obtain $\hat{\rho}$ by solving the generalized eigenvalue problem

$$U\rho = \delta V \rho \quad (3.14)$$

where δ denotes the eigenvalues of kernel alignment. Once, we find this optimum composite coefficient $\hat{\rho}$, which will be the eigenvector corresponding to the maximum eigenvalue δ, we can compute the composite data-dependent kernel matrix $K^s_{com}(\rho)$ according to Equation 3.9 by changing data clusters. That means that eigenvectors $\hat{\rho}$ for $U\rho = \delta V\rho$ provide the optimum coefficients for the composite kernel in Equation 3.9.

This composite kernel process provides an optimal data-dependent kernel. We can now proceed with the training of the multiple databases for the reclustered database in the subsequent section.

3.4.2 Multiple Databases Using Composite Kernel

We extend Section 3.4.1 for multiple databases after the composite kernels have been identified. Four basic kernels are considered to combine and represent six $K_{com}(\rho)$ shown in Fig. 3.6, as $K^1_{com}(\rho)$, $K^2_{com}(\rho)$, $K^3_{com}(\rho)$, and $K^5_{com}(\rho)$. We will assign each database (19 databases in our colonography experiment in Section 3.5) as one of six composite kernel cases. Database 1, for example, is labeled by kernel K^1_{com}, Databases 8 and 9 use kernel K^2_{com}. Then we assemble the database according the kernel labeled.

To assemble the databases, we further optimize combining coefficients by assembling databases. Our goal is to find a composite kernel that will best fit these newly assembled Databases K^s_{group}. As the composite kernel with the coefficients has been calculated in Section 3.4.1 starting from Equation 3.9, the desired calculation of the assembled databases uses the precalculated values K^s_{group}. Let us define the newly

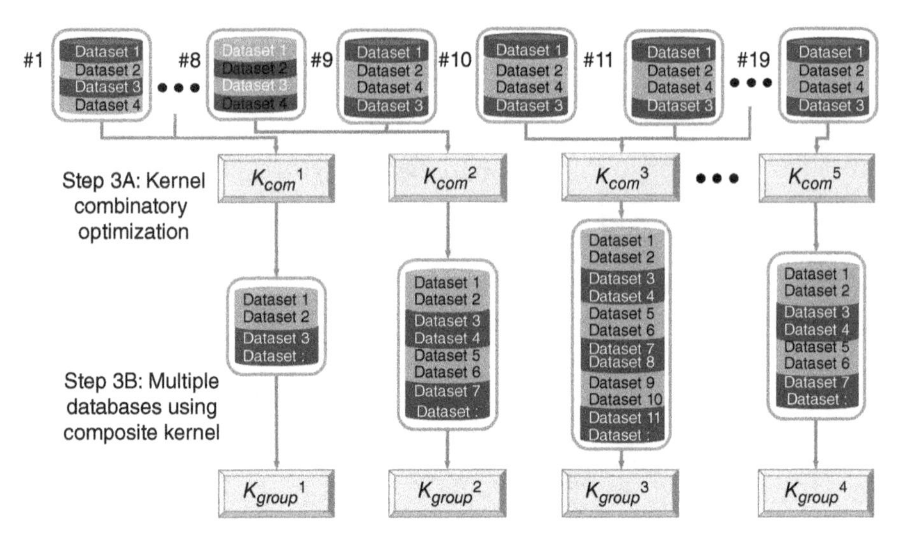

Figure 3.6 Group kernel feature analysis (GKFA). The first two steps 1 and 2 are the same as in Fig. 3.2. Step 3 of Fig. 3.2 is illustrated here for assembled databases through kernel choice in the composite kernel by Sections 3.4.1 and 3.4.2.

calculated group kernel K^s_{group} by the weighted sum of composite kernel previously calculated in Equation 3.14 in Section 3.4.1 as shown in Equation 3.15:

$$K^s_{group}(\hat{\rho}_{group}) = \sum_{g=1}^{Dg} \hat{\rho}_g (K^s_{com})_g \tag{3.15}$$

where K^s_{group} is the weighted sum of data-dependent optimized by assembling databases up to the number D_g, representing $K^s_{group} \cdot D_g$ is defined as the total database number assembled for grouping under the identical composite kernel K^s_{group}. We do not directly calculate K^s_{group} using Equation 3.15, as we already have calculated $\hat{\rho}$ and δ corresponding to K^s_{group} for the individual database in Section 3.4.1. We would like to estimate the optimal $\hat{\rho}_g$ shown in Equation 3.15 by the weighted sum of eigenvectors through the eigenvalues δ previously calculated in Equation 3.14:

$$\hat{\rho}_{group} = \frac{\sum_{g=1}^{D_g} \delta_g \hat{\rho}_g}{\sum_{g=1}^{D_g} \delta_g} \tag{3.16}$$

where the value δ_g denotes the eigenvalues of the database g in Equation 3.14, corresponding to the individual database in Section 3.4.1. The newly grouped kernel, corresponding to the largest eigenvalue δ_g, is used in the construction of an assembled database to yield the optimum classification accuracy. This entire process of Sections 3.4.1 and 3.4.2 is summarized in the flow chart in Fig. 3.7.

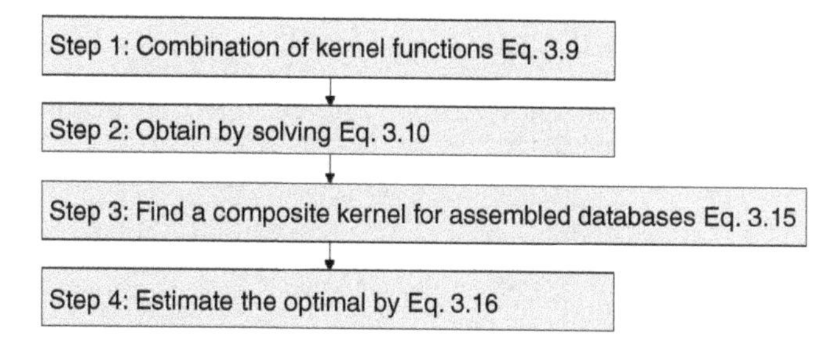

Figure 3.7 The overall GKFA steps of newly assembled databases.

While a single database is used for the approaches, the proposed GKFA approach uses more than one single database to improve the cancer classification performance as shown in the experimental results.

3.5 EXPERIMENTAL RESULTS

The experimental results consist of Sections 3.5.1–3.5.7.

3.5.1 *Cancer Databases*

We evaluated the performance of the proposed GKFA on the basis of a retrospectively established database of clinical cases obtained from several multicenter screening CTC trials [45–47]. The database consisted of 464 CTC cases that were obtained from 19 medical centers in the United States and Europe. Using the larger scale such as distributed datasets, confidentiality and auditability are the top issues addressed for the Health Insurance Portability and Accountability Act (HIPAA) through privacy, performance unpredictability, and scalable storage. This prototype study follows the HIPAA policy; the valuable diagnosis data contents on medical imaging platforms were closed with the data only available to system users, or very limitedly connected with outside users.

Our previously developed CAD scheme [48–50] was applied to these CTC volumes, which yielded a total of 3774 detections (polyp candidates) consisting of 136 true positive (TP) detections and 3638 false positive (FP) detections. Note that the supine and prone CTC volumes of a patient were treated as independent in the detection process. A volume of interest (VOI) of 963 pixels was placed at each candidate to cover the entire region of the candidate. The collection of the VOIs for all of the candidates consisted of the databases used for the performance evaluation as shown in Table 3.1. We applied up to 40%-fold cross-variation for testing the training data. Note that the training and testing data were separated for a more explicit form.

The proposed statistical analysis by use of GKFA was applied to the databases in Table 3.1, which showed that the CTC data were highly biased toward FPs (the average ratio between TP and FP is 1 : 26.6) due to the limited number of TPs caused by asymptomatic patient cohort. The proposed statistical analysis by use of GKFA is expected to compensate the lack of TPs by incorporating the multiple databases.

Table 3.1 Databases

	# Patients			# Database		
Databases	# TP patients	# FP patients	Total # patients	# TP	# FP	Total # database
1	5	30	35	12	155	167
2	3	29	32	5	217	222
3	3	10	13	7	213	220
4	3	27	30	5	206	211
5	7	35	42	12	196	208
6	3	25	28	6	198	204
7	3	24	27	6	208	214
8	1	28	29	4	200	204
9	3	17	20	8	190	198
10	3	22	25	7	198	205
11	4	23	27	8	181	189
12	3	29	32	4	191	195
13	2	11	13	8	208	216
14	3	27	30	5	188	193
15	3	15	18	7	147	154
16	3	5	8	8	221	229
17	3	12	15	7	169	176
18	2	25	27	12	169	181
19	2	11	13	5	183	188
Average	3.1	21.3	24.4	7.15	190.1	197.3

3.5.2 Optimal Selection of Data-Dependent Kernels

We used the method proposed in Section 3.3.1 to create four different data-dependent kernels and select the kernel that best fit the data and achieved optimum classification accuracy for each individual database. We determined the optimum kernel depending on the eigenvalue that yielded maximum separability. The performance measure used to evaluate the experimental results was defined as the ratio between the number of successfully classified polyps and the total number of polyps. Table 3.2 lists the eigenvalues λ and parameters of four kernels for each database calculated in Equation 3.5.

Table 3.2 shows the maximal eigenvalues corresponding to data-dependent kernel of an individual database. Among the four data-dependent kernels, the Sigmoid kernel was observed to achieve the best performance for most databases except for database 14. The kernel with the maximum eigenvalue is highlighted for each database in Table 3.2.

3.5.3 Kernel Combinatory Optimization

Once we find the kernel that yields the optimum eigenvalue, we select the two largest kernels to form the composite kernel in this study. For example, for database 1, we combined the Sigmoid and Gauss kernels to form the composite kernel. We observed that each database had different combinations for the composite kernels. We adopted the KFA algorithm to obtain the feature vectors and classified them using the k-nearest neighbor (K-NN) method with a metric of Euclidean distance.

Table 3.2 Eigenvalues of four kernels for offline databases

Databases	The first kernel	The second kernel
1	Sigmoid kernel $\lambda = \boldsymbol{107}$, $d = 3.334 * 10^{-4}$, Offset = 0	Gauss kernel $\lambda = \boldsymbol{3.93}$, $\sigma = 0.7$
2	Sigmoid kernel $\lambda = \boldsymbol{31.6}$, $d = 2.223 * 10^{-4}$, Offset = 0	Polynomial kernel $\lambda = \boldsymbol{2.41}$, $d = 5$, Offset = 0.1
3	Sigmoid kernel $\lambda = \boldsymbol{96.7}$, $d = 1.0 * 10^{-7}$, Offset = 0	Polynomial kernel $\lambda = \boldsymbol{5.59}$, $d = 1$, Offset = 0.1
4	Sigmoid Kernel $\lambda = \boldsymbol{105}$, $d = 1.112 * 10^{-4}$, Offset = 0	Linear Kernel $\lambda = \boldsymbol{9.21}$, $d = 3$
5	Sigmoid kernel $\lambda = \boldsymbol{328}$, $d = 3.334 * 10^{-4}$, Offset = 0	Gauss kernel $\lambda = \boldsymbol{4.44}$, $\sigma = 0.1$
6	Sigmoid kernel $\lambda = \boldsymbol{80.1}$, $d = 2.223 * 10^{-4}$, Offset = 0	Gauss kernel $\lambda = \boldsymbol{9.82}$, $\sigma = 0.8$
7	Sigmoid kernel $\lambda = \boldsymbol{38.3}$, $d = 2.223 * 10^{-4}$, Offset = 0	Linear kernel $\lambda = \boldsymbol{17.1}$, $d = 5$
8	Sigmoid kernel $\lambda = \boldsymbol{127}$, $d = 5.556 * 10^{-4}$, Offset = 0	Gauss kernel $\lambda = \boldsymbol{8.31}$, $\sigma = 0.1$
9	Sigmoid kernel $\lambda = \boldsymbol{35.9}$, $d = 1.112 * 10^{-4}$, Offset = 0	Gauss kernel $\lambda = \boldsymbol{20.5}$, $\sigma = 0.9$
10	Sigmoid kernel $\lambda = \boldsymbol{18.6}$, $d = 1.112 * 10^{-4}$, Offset = 0	Linear kernel $\lambda = \boldsymbol{2.28}$, $d = 3$
11	Sigmoid kernel $\lambda = \boldsymbol{52}$, $d = 2.223 * 10^{-4}$, Offset = 0	Gauss kernel $\lambda = \boldsymbol{2.53}$, $\sigma = 0.1$
12	Sigmoid kernel $\lambda = \boldsymbol{88.9}$, $d = 4.445 * 10^{-4}$, Offset = 0	Gauss kernel $\lambda = \boldsymbol{14.8}$, $\sigma = 0.6$
13	Sigmoid kernel $\lambda = \boldsymbol{40.8}$, $d = 2.223 * 10^{-4}$, Offset = 0	Gauss kernel $\lambda = \boldsymbol{1.78}$, $\sigma = 0.1$
14	Polynomial kernel $\lambda = 29.6$, $d = 1$, Offset = 0.1	Sigmoid kernel $\lambda = 6.3$, $d = 0.000001$, Offset = 0
15	Sigmoid kernel $\lambda = 280$, $d = 3.334 * 10^{-4}$, Offset = 0	Gauss kernel $\lambda = 11.0$, $\sigma = 0.7$
16	Sigmoid kernel $\lambda = 48.1$, $d = 3.334 * 10^{-4}$, Offset = 0	Gauss kernel $\lambda = 1.82$, $\sigma = 0.1$
17	Sigmoid kernel $\lambda = 89.0$, $d = 3.334 * 10^{-4}$, Offset = 0	Polynomial kernel $\lambda = 1.28$, $d = 7$, Offset = 0.1
18	Sigmoid kernel $\lambda = \boldsymbol{179}$, $d = 2.223 * 10^{-4}$, Offset = 0	Gauss kernel $\lambda = 1.38$, $\sigma = 0.1$
19	Sigmoid kernel $\lambda = 46.2$, $d = 3.334 * 10^{-4}$, Offset = 0	Gauss kernel $\lambda = 1.35$, $\sigma = 0.1$

Table 3.3 The value of $\hat{\rho}$ for each of the composite kernels

Databases	Two most dominant kernels	ρ_1	ρ_2	K-NN(k)	Performance, %
1	Sigmoid and Gauss	0.73	0.27	1	98.00
2	Sigmoid and poly	0.27	0.73	8	90.00
3	Sigmoid and poly	0.68	0.32	7	86.27
4	Sigmoid and linear	0.94	0.06	3	94.23
5	Sigmoid and Gauss	0.30	0.70	1	92.16
6	Sigmoid and Gauss	0.72	0.28	1	97.78
7	Sigmoid and linear	0.27	0.73	3	90.20
8	Sigmoid and Gauss	0.31	0.69	1	96.00
9	Sigmoid and Gauss	0.65	0.35	1	92.16
10	Sigmoid and linear	0.27	0.73	1	92.00
11	Sigmoid and Gauss	0.28	0.72	1	98.04
12	Sigmoid and Gauss	0.24	0.76	3	94.12
13	Sigmoid and Gauss	0.27	0.73	3	92.16
14	Sigmoid and poly	0.43	0.57	5	90.38
15	Sigmoid and Gauss	0.73	0.27	1	95.35
16	Sigmoid and Gauss	0.27	0.73	1	97.96
17	Sigmoid and poly	0.31	0.69	1	91.84
18	Sigmoid and Gauss	0.74	0.26	1	94.00
19	Sigmoid and Gauss	0.27	0.73	4	90.91

Table 3.3 shows how the two kernel functions are combined according to the composite coefficients listed in the table. These composite coefficients were obtained in Section 3.4.2. For all of the databases, the most dominant kernels kept varying, and the second most dominant kernel was the Sigmoid kernel. As a result, the contribution of the Sigmoid kernel was lower when compared to other kernels in forming a composite kernel.

3.5.4 Composite Kernel for Multiple Databases

We used the method proposed in Section 3.4.2 to obtain the group kernel by the weighted sum of the composite kernels, then assembled 19 individual databases according to kernel type. As for the Sigmoid and Gauss group kernel, we sorted 12 databases in order as: $\lambda^9_* < \lambda^{13}_* < \lambda^{11}_* < \lambda^{16}_* < \lambda^{19}_* < \lambda^6_* < \lambda^{12}_* < \lambda^1_* < \lambda^8_* < \lambda^{18}_* < \lambda^{15}_* < \lambda^5_*$, and then divided them into three assembled databases by maximal eigenvalue in Equation 3.4. We applied the GKFA method with the K-NN classifier to the assembled databases, using the classification performance and parameters shown in Table 3.4.

Table 3.4 shows the way to assemble the database by group kernel. Performance can be compared with the values in Table 3.3. It shows that most databases can be categorized into the Sigmoid and Gauss group kernel. We divided 12 databases into three assembled databases (CASE1, CASE2, and CASE3) according to order of eigenvalue. We observed classification performance of 98.49% (average) over KFA with individual Database of 95.22%. The second database assembled was the Sigmoid and Poly group kernel. A classification rate of 95.33% was achieved, which outperformed

Table 3.4 GKFA for assembled database

Kernel type		Database assembled	First kernel	Second kernel	Performance, %
Sigmoid and Gauss	CASE1	Database9	*Sigmoid*	*Gauss*	*97.54*
		Database13	$d = 6.67 * 10^{-4}$	$\sigma = 0.4$	
		Database11	Offset = 0	$\rho_2 = 0.51$	
		Database16	$\rho_1 = 0.49$		
	CASE2	Database19	*Sigmoid*	*Gauss*	*100.00*
		Database6	$d = 8.89 * 10^{-4}$	$\sigma = 0.2$	
		Database12	Offset = 0	$\rho_2 = 0.12$	
		Database1	$\rho_1 = 1.00$		
	CASE3	Database8	*Sigmoid*	*Gauss*	*97.94*
		Database18	$d = 7.78 * 10^{-4}$	$\sigma = 0.3$	
		Database15	Offset = 0	$\rho_2 = 0.05$	
		Database5	$\rho_1 = 0.95$		
Sigmoid and poly		Database2	*Sigmoid*	*Linear*	*95.33*
		Database3	$d = 3.34 * 10^{-4}$	$d = 3$,	
		Database14	Offset = 0	Offset = 0.1	
		Database17	$\rho_1 = 0.71$	$\rho_2 = 0.20$	
Sigmoid and linear		Database4	*Sigmoid*	*Linear*	*97.35*
		Database7	$d = 3.34 * 10^{-4}$	$d = 13$	
		Database10	Offset = 0	$\rho_2 = 0.12$	
			$\rho_1 = 1.00$		

KFA with individual databases 2, 3, 14, and 17 by an average of 89.62%. The last database assembled was the Sigmoid and linear group. We obtained a 97.35% performance rate, which was 5.2% higher than KFA for databases 4, 7, and 10. So we can conclude from the comparison of Tables 3.3 and 3.4, that the classification performance can be improved by using GKFA for the assembled databases over KFA for a single database.

3.5.5 K-NN Classification Evaluation with ROC

We demonstrate the advantage of GKFA in terms of the receiver operating characteristic (ROC) in a more specific way by comparing GKFA for assembled databases to KFA for a single database. We evaluated the classification accuracy shown in Table 3.3, and ROC using the sensitivity and specificity criteria as statistical measures of the performance. The true positive rate (TPR) defines a diagnostic test performance for classifying positive cases correctly among all positive instances available during the test. The false positive rate (FPR) determines how many incorrect positive results occur among all negative samples available during the test as follows [51]:

$$\text{TPR} = \frac{\text{True Positives (TP)}}{\text{True Positives (TP)} + \text{False Negatives (FN)}} \tag{3.17}$$

$$\text{FPR} = \frac{\text{False Positives (FP)}}{\text{False Positives (FP)} + \text{True Negatives (TN)}} \tag{3.18}$$

Table 3.5 Performance with Sigmoid and Gauss group kernel (CASE 1)

Performance metrics	Database 9	Database 11	Database 13	Database 16	Assembling databases
AUC of ROC	0.94	0.97	0.95	0.97	0.99
Rand Index	0.842	0.958	0.839	0.949	0.972

Table 3.6 Performance with Sigmoid and Gauss group kernel (CASE2)

Performance metrics	Database 1	Database 6	Database 12	Database 19	Assembling databases
AUC of ROC	0.97	0.82	0.93	0.38	0.99
Rand Index	0.953	0.943	0.857	0.835	0.982

Table 3.7 Performance with Sigmoid and Gauss group kernel (CASE3)

Performance metrics	Database 5	Database 8	Database 15	Database 18	Assembling databases
AUC of ROC	0.85	0.50	0.79	0.99	0.98
Rand Index	0.857	0.971	0.754	0.989	0.953

The classification performance using ROC is evaluated by the area under the curve (AUC) by calculating the integral of ROC plot in the range from 0 to 1. We also evaluated the classification performance with the Rand Index. Table 3.8–3.10 show those performance corresponding to the key type of Sigmoid and Gauss shown in CASE 1, CASE2, and CASE3 of Table 3.4.

Tables 3.5–3.7 show that the classification results for the Sigmoid and Gauss group kernels. We compared the assembled databases performance by AUC and Rand Index, with a single database (databases 9, 11, 13, and 16) in Table 3.5, a single database (databases 1, 6, 12, and 19) in Table 3.6, and a single database (databases 5, 8, 15, and 18) in Table 3.7, respectively. In Table 3.5, GKFA for the assembled database outperformed the KFA for a single database by achieving a higher TPR and a lower FPR. In Table 3.6, although database 1 had a smaller FPR, GKFA for the assembled database had better result. In Table 3.7, the performance of GKFA for the assembled database was not as good as KFA for database 18, but it performed better than the other three databases.

Table 3.8 shows that KFA with database 17 had the best performance and the second best one was the performance of GKFA for the assembled database. Regarding KFA for databases 2, 3, and 14, FPR was a bit high, and TPR was not good enough at the same time. In Sigmoid and poly group kernel experiment, GKFA for the assembled databases was not the best one, but better than KFA for the rest of the single databases.

Table 3.8 Performance with Sigmoid and poly group kernel

Performance metrics	Database 2	Database 3	Database 14	Database 17	Assembling databases
AUC of ROC	0.32	0.12	0.87	0.99	0.85
Rand Index	0.514	0.701	0.695	0.989	0.905

Table 3.9 Performance with Sigmoid and linear group kernel

Performance metrics	Database 4	Database 7	Database 100	Assembling databases
AUC of ROC	0.40	0.17	0.57	0.97
Rand Index	0.963	0.945	0.962	0.936

Table 3.10 comparison of different classification methods

Classifier method	Classifier parameters	Performance, %
RNN	Number of iterations given 2500	90.00
BNN	Number of iterations given1500	82.11
SVM	Gauss kernel, Gaussian width 0.001, slack 0.1	87.37
DT	Incorrectly assigned samples at a node 10%	96.32
K-NN	***K* number 1**	**97.94**

In Table 3.9, we can see that GKFA for the assembled database outperformed KFA for databases 4, 7, and 10, although database 10 had small FPR, but the maximal TPR only reached 0.58. On the other hand, GKFA for the assembled databases reached 0.98, although FPR was bigger than database 10. In most cases, GKFA for the assembled databases had the advantage over KFA by an average of 22.2% in TPR and 3.8% in FPR for a single database with respect to FPR and TPR.

3.5.6 Comparison of Results with Other Studies on Colonography

In this section, we compared results for the Sigmoid and Gauss group kernels for four databases (databases 1, 6, 12, and 19) and the assembled databases with different classifiers, which include RBF neural networks (RNN), back-propagation neural network (BNN), support vector machine (SVM), decision tree (DT), and K-NN method [52]. Classification performance is shown in Table 3.5 through a MATLAB® implementation, and in Table 3.9 we show the resulting ROC curves. In Table 3.10, we can see that the K-NN method yielded the best performance followed by DT, RNN, SVM, BNN methods in descending order. Each classifier is listed with its corresponding parameters. The construction of NN is three-layer network with fixed increment single-sample perceptron for BNN, and RNN has radial basis function in the middle layer.

In Table 3.11, we can also see that the assembled databases with the K-NN method achieved the best performance, followed by DT with less TPR. So we can conclude

Table 3.11 Classification performance with different methods

Performance metrics	RNN	BNN	SVM	DT	KNN
AUC of ROC	0.82	0.35	0.54	0.89	0.99
Rand Index	0.685	0.748	0.808	0.948	0.981

Table 3.12 Classification accuracy to compare other methods

Studies	Performance, %	Datasets	Methods
CAD [6]	Colonic polyps: 100 Lung: 90	Colonic polyps and lung modules in helical CT	Surface normal overlap method
Polyp segmentation [15]	*91.8*	CTC polyp	Fuzzy clustering and deformable models
CTC detection [53]	*94*	CTC polyps	Logistic regression
CTC cleansing [54]	*94.6*	Phantom and CTC polyps	Mosaic decomposition
Multiple CTC	***97.63***	**CTC polyps**	**Group kernel feature analysis (GKFA)**

that the K-NN method was the most appropriate classifier for the distributed medical imaging databases with group kernel feature analysis in the experiment.

Other existing studies investigated the classification of CTC showed the comparative experiments results [6, 15, 53, 54]. On the basis of the proposed GKFA method applied to multiple CTC databases, the classification performance was listed in Table 3.12.

Note that in this comparison table, the datasets of CTC were different from each other, so the actual classification performance ratio is not suitable enough to compare each other by different methods reported by investigators using different datasets. At least, the proposed GKFA classification yielded an overall performance of 97.63% in average with the standard deviation of 2.77%; thus the proposed classifier would potentially outperform existing colonography studies [6, 15, 53, 54].

This is the first study to demonstrate whether multiple databases may improve the classification performance. The future step is to increase the size of databases and institutions. The current size of 464 patients total from 19 databases only has 405 normal cases and 59 abnormal cases. Furthermore, one identical CAD system may be extended into other CAD systems such as most distributed computing handle other modal-platforms through an interfacing module among multiple CAD databases.

3.5.7 Computational speed and scalability evaluation of GKFA

The computational efficiency of the proposed GKFA method was evaluated by comparing its run time with KPCA and KFA for the selected datasets. The algorithms have been implemented in MATLAB R2007b using the Statistical Pattern Recognition Toolbox for the Gram matrix calculation and kernel projection. The processor

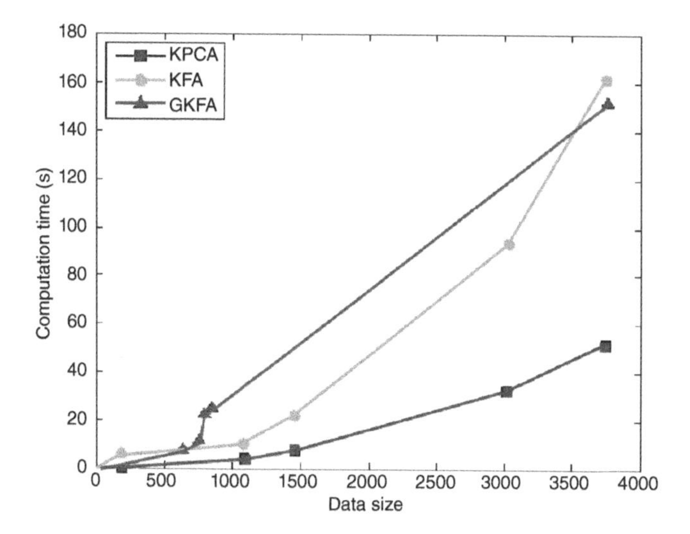

Figure 3.8 The computation time comparison between KPCA, KFA, and GKFA as the data size increases.

was a 3.2 GHz Intel® Pentium 4 CPU with 3 GB of random access memory (RAM). Run time was determined using the cputime command. For each algorithm, computation time increases with increasing training data size (n), as expected. All three methods required the computation of a Gram matrix whose size increases as the data size increased. The results from Fig. 3.8 clearly indicated that GKFA and KFA required more computational time than KPCA because the composite data-dependent kernels needed further calculations of a Gram matrix and optimization of coefficient parameters. These overhead computations mainly caused the increase in run time.

Typically, the overall scalability is estimated by the computational complexity. If the KPCA algorithm contains an eigenvalue problem of rank n, the computational complexity of KPCA is $O(n^3)$. In addition, each resulting eigenvector is represented as a linear combination of n terms; the l features depend on n image vectors of X_n. Thus, all data contained in X_n must be retained, which is computationally cumbersome and unacceptable for our distributed applications. If the KFA algorithm contains an eigenvalue problem of rank n, the computational complexity of KFA is expected to be $O(n^2)$. If the GKFA algorithm contains an distributed problem of D_g databases, the computational complexity of GKFA is expected to be $O(D_g n_g{}^2)$.

3.6 CONCLUSIONS

In this study, we proposed a new framework for health informatics: CAD of cancer polyps using distributed colonography, where distributed databases from multiple hospitals are considered for participation. We showed how to merge the information-centric character and node-centric physical world connectivity to develop smart healthcare systems. This is the first pilot study on the evaluation of how the proposed machine-supported diagnosis can specifically handle multiple colonography databases.

The size of TP data is usually small due to the screening test at one specific institution. The CAD algorithm and clinicians both can observe all of the potential cases in the proposed distributed platform. For handling multiple CTC databases, the proposed method, called group kernel feature analysis (GKFA), was developed in this study. When GKFA is used for assembled databases, it can achieve an average improvement of 22.2% in TPR and 3.8% in FPR compared to single databases. GKFA has the potential to play as a core classifier tool in the distributed computing framework for a model-based CAD scheme, which will yield high detection performance of polyps using multiple institutions databases. Successful development of CAD in the distributed computing environment may advance the clinical implementation of cancer screening and promote the early diagnosis of colon cancer, so that the proposed scheme can make CTC a viable option for screening large patient populations, resulting in early detection of colon cancer and leading to reduced mortality due to colon cancer.

REFERENCES

1. S. Winawer, R. Fletcher, D. Rex, J. Bond, R. Burt, J. Ferrucci, T. Ganiats, T. Levin, S. Woolf, D. Johnson, L. Kirk, S. Litin, and C. Simmang, "Colorectal cancer screening and surveillance: Clinical guidelines and rationale-update based on new evidence," *Gastroenterology*, vol. 124, pp. 544–60, 2003.
2. J. G. Fletcher, F. Booya, C. D. Johnson, and D. Ahlquist, "CT colonography: Unraveling the twists and turns," *Curr. Opin. Gastroenterol.*, vol. 21, pp. 90–98, 2005.
3. Z. B. Xu, M. W. Dai, and D. Y. Meng, "Fast and efficient strategies for model selection of Gaussian support vector machine," *IEEE Trans. Syst. Man Cyb. B*, vol. 39, pp. 1292–1207, 2009.
4. J. Näppi, H. Frimmel, A. H. Dachman, and H. Yoshida, "A new high-performance CAD scheme for the detection of polyps in CT colonography," in *Med Imaging 2004: Image Processing*, San Diego, CA, February 14, pp. 839–848, 2004.
5. G. Kiss, J. Van Cleynenbreugel, M. Thomeer, P. Suetens, and G. Marchal, "Computer-aided diagnosis in virtual colonography via combination of surface normal and sphere fitting methods," *Eur. Radiol.*, vol. 12, pp. 77–81, Jan. 2002.
6. D. S. Paik, C. F. Beaulieu, G. D. Rubin, B. Acar, R. B. Jeffrey, Jr., J. Yee, J. Dey, and S. Napel, "Surface normal overlap: A computer-aided detection algorithm with application to colonic polyps and lung nodules in helical CT," *IEEE Trans. Med Imaging*, vol. 23, pp. 661–675, 2004.
7. A. K. Jerebko, R. M. Summers, J. D. Malley, M. Franaszek, and C. D. Johnson, "Computer-assisted detection of colonic polyps with CT colonography using neural networks and binary classification trees," *Med. Phys.*, vol. 30, pp. 52–60, 2003.
8. A. K. Jerebko, J. D. Malley, M. Franaszek, and R. M. Summers, "Multiple neural network classification scheme for detection of colonic polyps in CT colonography databases," *Acad. Radiol.*, vol. 10, pp. 154–160, 2003.

9. A. K. Jerebko, J. D. Malley, M. Franaszek, and R. M. Summers, "Support vector machines committee classification method for computer-aided polyp detection in CT colonography," *Acad. Radiol.*, vol.12, pp. 479–486, 2005.
10. M. Doumpos, C. Zopounidis, and V. Golfinopoulou, "Additive support vector machines for pattern classification," *IEEE Trans. Syst. Man Cyb. B*, vol. 37, pp. 540–550, 2007.
11. R. M. Summers, M. Franaszek, M. T. Miller, P. J. Pickhardt, J. R. Choi, and W. R. Schindler, "Computer-aided detection of polyps on oral contrast-enhanced CT colonography," *AJR Am. J. Roentgenol.*, vol. 184, pp. 105–108, 2005.
12. J. Nappi, H. Frimmel, and H. Yoshida, "Virtual endoscopic visualization of the colon by shape-scale signatures," *IEEE Trans. Inform. Tech. Biomed.*, vol. 9, pp. 120–131, 2005.
13. L. Tong-Yee, L. Ping-Hsien, L. Chao-Hung, S. Yung-Nien, and L. Xi-Zhang, "Interactive 3-D virtual colonoscopy system," *IEEE Trans. Inform. Tech. Biomed.*, vol. 3, pp. 139–150, 1999.
14. T. T. Chang, J. Feng, H. W. Liu, and H. Ip, "Clustered microcalcification detection based on a multiple kernel support vector machine with grouped features," in *Proc 19th Int. Conf. Pattern Recognition*, December, 2008, Tampa, FL, vol. 8, pp. 1–4.
15. J. H. Yao, M. Miller, M. Franaszek, and R.M. Summers, "Colonic polyp segmentation in CT colonography-based on fuzzy clustering and deformable models," *IEEE Trans. Med. Imaging*, vol. 23, pp. 1344–1356, 2004.
16. M. Xu, P. M. Thompson, and A. W. Toga, "Adaptive reproducing kernel particle method for extraction of the cortical surface," *IEEE Trans. Med Imaging*, vol. 25, no. 6, pp. 755–762, 2006.
17. I. E. A. Rueda, F. A. Arciniegas, and M. J. Embrechts, "Bayesian separation with sparsity promotion in perceptual wavelet domain for speech enhancement and hybrid speech recognition," *IEEE Trans. Syst. Man Cyb. A*, vol. 34, pp. 387–398, 2004.
18. W. Zheng, C. Zou, and L. Zhao, "An improved algorithm for kernel principal component analysis," *Neural Process Lett.*, pp. 49–56, 2005.
19. J. Kivinen, A. J. Smola, and R. C. Williamson, "Online learning with kernels," *IEEE Trans. Signal Process.*, vol. 52, no. 8, pp. 2165–2176, 2004.
20. S. Ozawa, S. Pang, and N. Kasabov, "Incremental learning of chunk data for online pattern classification systems," *IEEE Trans. Neural Netw.*, vol. 19, no. 6, pp. 1061–1074, 2008.
21. H. T. Zhao, P. C. Yuen, and J. T. Kwok, "A novel incremental principal component analysis and its application for face recognition," *IEEE Trans. Syst. Man Cyb. B*, vol. 36, no. 4, pp. 873–886, 2006.
22. Y. M. Li, "On incremental and robust subspace learning," *Pattern Recogn.*, vol. 37, no. 7, pp. 1509–1518, 2004.
23. Y. Kim, "Incremental principal component analysis for image processing," *Opt. Lett.*, vol. 32, no. 1, pp. 32–34, 2007.
24. B. J. Kim, and I. K. Kim, "Incremental nonlinear PCA for classification," in *Proc. Knowledge Discovery in Databases*, Pisa, Italy, September 20–24, vol. 3202, pp. 291–300, 2004.
25. B. J. Kim, J. Y. Shim, C. H. Hwang, I. K. Kim, and J. H. Song. "Incremental feature extraction based on empirical kernel map," *Found. Intell. Syst.*, vol. 2871, pp. 440–444, 2003.
26. L. Hoegaerts, L. De Lathauwer, I. Goethals, J. A. K. Suykens, J. Vandewalle, and B. De Moor. "Efficiently updating and tracking the dominant kernel principal components," *Neural Netw.*, vol. 20, no. 2, pp. 220–229, 2007.

27. T. J. Chin, and D. Suter. "Incremental kernel principal component analysis," *IEEE Trans. Image Process.*, vol. 16, no. 6, pp. 1662–1674, 2007.
28. B. Schökopf and A. J. Smola, "Learning with kernels: Support vector machines, regularization, optimization, and beyond," *Adaptive computation and machine learning*. Cambridge, MA: MIT Press, 2002.
29. H. Fröhlich, O. Chapelle, and B. Scholkopf, "Feature selection for support vector machines by means of genetic algorithm," in *Proc. 15th IEEE Int. Conf. Tools with Artificial Intelligence*, Washington DC, December 3–5, pp. 142–148, 2003.
30. V. N. Vapnik, *The nature of statistical learning theory*, 2nd ed. New York: Springer, 2000.
31. T. Damoulas and M. A. Girolami, "Probabilistic multi-class multi-kernel learning: On protein fold recognition and remote homology detection," *Bioinformatics*, vol. 24, no. 10, pp. 1264–1270, 2008.
32. X. W. Chen, "Gene selection for cancer classification using bootstrapped genetic algorithms and support vector machines," in *Proc. IEEE Int. Conf. Computational Systems, Bioinformatics Conference*, Stanford, CA, August 11–14, pp. 504–505, 2003.
33. C. Park and S. B. Cho, "Genetic search for optimal ensemble of feature-classifier pairs in DNA gene expression profiles," in *Proc. Int. Joint Conf. Neural Networks*, Portland, OR, July 20–24, vol. 3, pp. 1702–1707, 2003.
34. F. A. Sadjadi, "Polarimetric radar target classification using support vector machines," *Opt. Eng.*, vol. 47, no. 4, pp. 046201, 2008.
35. X. Jiang, R. Snapp, Y. Motai, and X. Zhu, "Accelerated kernel feature analysis," in *Proc. IEEE Computer Society Conference on Computer Vision and Pattern Recognition*, New York, NY, June 17–22, pp. 109–116, 2006.
36. L. Jayawardhana, Y. Motai, and A. Docef, "Computer-aided detection of polyps in CT colonography: On-line versus off-line accelerated kernel feature analysis," Special Issue on Processing and Analysis of High-Dimensional Masses of Image and Signal Data, *Signal Process.*, pp. 1–12, 2009.
37. A. K. Jerebko, R. M. Summers, J. D. Malley, M. Franaszek, and C. D. Johnson, "Computer-assisted detection of colonic polyps with CT colonography using neural networks and binary classification trees," *Med. Phys.*, vol. 30, pp. 52–60, 2003.
38. H. Xiong, Y. Zhang, and X. W. Chen, "Data-dependent kernel machines for microarray data classification," *IEEE/ACM Trans. Comput. Biol. Bioinform.*, vol. 4, no. 4, pp. 583–595, 2007.
39. Y. Motai and H. Yoshida, "Principal composite kernel feature analysis: Data-dependent kernel approach," *IEEE Trans. Knowl. Data Eng.*, vol. 25, no. 8, pp. 1863–1875, 2013.
40. N. Cristianini, J. Kandola, A. Elisseeff, and J. Shawe-Taylor, "On kernel target alignment," *Proc. Neural Inform. Process. Syst.*, pp. 367–373, 2001.
41. P. J. Pickhardt, J. R. Choi, I. Hwang, J. A. Butler, M. L. Puckett, H. A. Hildebrandt, R. K. Wong, P. A. Nugent, P. A. Mysliwiec, and W. R. Schindler, "Computed tomographic virtual colonoscopy to screen for colorectal neoplasia in asymptomatic adults," *N. Engl. J. Med.*, vol. 349, pp. 2191–2200, 2003.
42. D. C. Rockey, E. Paulson, D. Niedzwiecki, W. Davis, H. B. Bosworth, L. Sanders, J. Yee, J. Henderson, P. Hatten, S. Burdick, A. Sanyal, D. T. Rubin, M. Sterling, G. Akerkar, M. S. Bhutani, K. Binmoeller, J. Garvie, E. J. Bini, K. McQuaid, W. L. Foster, W. M. Thompson, A. Dachman, and R. Halvorsen, "Analysis of air contrast barium enema, computed tomographic colonography, and colonoscopy: Prospective comparison," *Lancet*, vol. 365, pp. 305–11, 2005.

43. D. Regge, C. Laudi, G. Galatola, P. Della Monica, L. Bonelli, G. Angelelli, R. Asnaghi, B. Barbaro, C. Bartolozzi, D. Bielen, L. Boni, C. Borghi, P. Bruzzi, M. C. Cassinis, M. Galia, T. M. Gallo, A. Grasso, C. Hassan, A. Laghi, M. C. Martina, E. Neri, C. Senore, G. Simonetti, S. Venturini, and G. Gandini, "Diagnostic accuracy of computed tomographic colonography for the detection of advanced neoplasia in individuals at increased risk of colorectal cancer," *JAMA*, vol. 301, no. 23, pp. 2453–2461, 2009.
44. H. Yoshida and J. Nappi, "Three-dimensional computer-aided diagnosis scheme for detection of colonic polyps," *IEEE Trans. Med. Imaging*, vol. 20, pp. 1261–1274, 2001.
45. J. Näppi and H. Yoshida, "Fully automated three-dimensional detection of polyps in fecal-tagging CT colonography," *Acad. Radiol.*, vol. 14, pp. 287–300, 2007.
46. H. Yoshida, J. Näppi, and W. Cai, "Computer-aided detection of polyps in CT colonography: Performance evaluation in comparison with human readers based on large multicenter clinical trial cases," in *Proc 6th IEEE Int. Symp. Biomedical Imaging*, Boston, MA, June 28–July 1, pp. 919–922, 2009.
47. T. Fawcett, "An introduction to ROC analysis," *Pattern Recogn. Lett.*, vol. 27, no. 8, pp. 861–874, 2006.
48. D. G. Stork and E. Yom-Tov, *Computer manual in MATLAB to accompany pattern classification*. 2nd ed., Hoboken, NJ: John Wiley & Sons, 2004.
49. V. F. van Ravesteijn, C. van Wijk, F. M. Vos, R. Truyen, J. F. Peters, J. Stoker, and L. J. van Vliet, "Computer-aided detection of polyps in CT colonography using logistic regression," *IEEE Trans. Med. Imaging*, vol. 29, pp. 120–131, 2010.
50. C. L. Cai, J. G. Lee, M. E. Zalis, and H. Yoshida, "Mosaic decomposition: An electronic cleansing method for inhomogeneously tagged regions in noncathartic CT colonography," *IEEE Trans. Med. Imaging*, vol. 30, pp. 559–570, 2011.
51. V. Taimouri, L. Xin, L. Zhaoqiang, L. Chang, P. Darshan, H. Jing, "Colon segmentation for prepless virtual colonoscopy," *IEEE Trans. Inform. Tech. Biomed.*, vol. 15, pp. 709–715, 2011.
52. C. Hu, Y. Chang, R. Feris, and M. Turk, "Manifold based analysis of facial expression," in *Proc. Computer Vision and Pattern Recognition Workshop*, Washington DC, June 27–July 2, vol. 27, pp. 81–85, 2004.
53. H. Yoshida and A. H. Dachman, "CAD techniques, challenges and controversies in computed tomographic colonography," *Abdom Imaging*, vol. 30, pp. 26–41, 2005.
54. R. M. Summers, C. F. Beaulieu, L. M. Pusanik, J. D. Malley, R. B. Jeffrey, Jr., D. I. Glazer, and S. Napel, "Automated polyp detector for CT colonography: Feasibility study," *Radiology*, vol. 216, pp. 284–290, 2000.

4

ONLINE KERNEL ANALYSIS

4.1 INTRODUCTION

A major limiting factor for screening colon cancer is the physicians' interpretation time [1–3]. This time-consuming process requires 20–25 min of reading images on the basis of ACRIN trial [4]. The image databases used in this study are cancer 4D datasets obtained from clinical colon screening in its applicability to computer-assisted detection (CAD) as shown in Fig. 4.1 [5–8]. The radiologist workload can be reduced, and diagnostic accuracy of the computer tomography colonography (CTC) [9–12] can be improved, with less inter-radiologist and interinstitutional variability in diagnostic accuracy. We seek to improve CAD automatic classification performance in a more realistic setting; everyday, additional CTC datasets of new patients are added into a pre-established database. We propose an online data association for the increasing amount of CTC databases for the larger number of patients. Colonography data and knowledge of radiologists are integrated so that the CAD's performance can be improved by exploiting information from larger database with more patients, diverse populations, and disparities [9, 13]. The concept of online learning has been discussed as an efficient method for nonlinear and online data analysis [14, 15].

Kernel analysis appears to be useful for medical imaging; however, there is no evidence that an automated diagnosis of cancer detection and classification is effective for the datasets and the increasing size of heterogeneous characteristics [16, 17]. The recent success of kernel methods is capitalizing on broad pattern classification problems [16, 18–21], especially in nonlinear classification and clustering. In this active approach of kernel methods, we have developed a faster feature analysis technique called accelerated kernel feature analysis (AKFA) [22] and have extended it to Composite Kernel Analysis, called principal composite kernel feature analysis

Data-Variant Kernel Analysis, First Edition. Yuichi Motai.

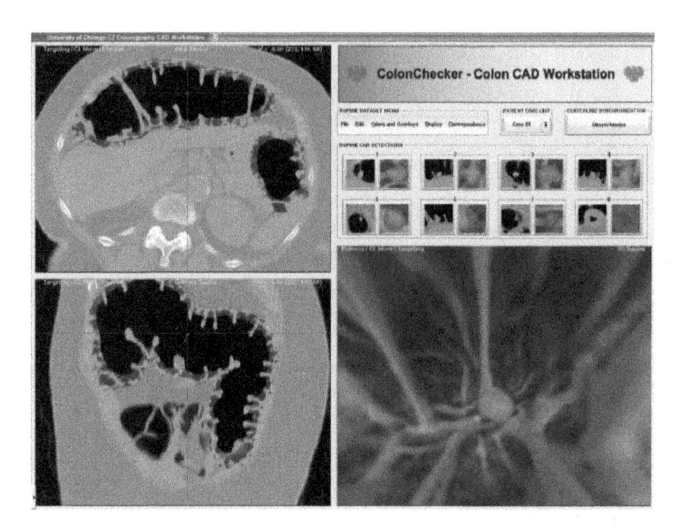

Figure 4.1 Developed CAD for CTC at 3D imaging group at Harvard Medical School [5].

(PC-KFA) [21]. These methods map the original, raw feature space into a higher dimensional feature space. Such a high-dimensional feature space is expected to have a greater classification power than that of the original feature space, as suggested by the Vapnik–Chervonenkis theory [23]. The algorithm extracts texture-based features from the polyp candidates generated by a shape-based CAD scheme. In this work, we propose a new online method, called online PC-KFA, designed for heterogeneous big data associations (HBDA).

The contribution is that this study tests real CTC data with the potential of making computer tomography (CT) a viable option for screening large patient populations, resulting in early detection of colon cancer and leading to reduced mortality rate. When handling mass data of clinical colon screening, there is a limitation of physician resources due to the human's time, accuracy, and cost. The extensive computational approach has a potential to improve overall classification performance, uniformity, and speed of CAD. The main contributions herein are to introduce a scale-free framework to manage the diversity and complexity of the much larger CTC databases. Improvements that could reduce reading time would go a long way toward making CTC a more acceptable technique by improving CAD performance [16, 18]. The inherent heterogeneity challenges our capacity to capitalize on the unity of the discipline—whether seen from computational, clinical, or resource perspectives. There are several essential issues that need to be addressed to make larger data-driven CAD feasible such as data confidentiality, auditability, and scalable storage.

The significance is that this is the first study to handle large CTC data by incorporating online data association. We are developing a new research solution for the problems of massive database expansions in scale, diversity, and complexity. The HBDA is ideal for solving heterogeneous data problems because the data acquired over a long period of time can sometimes be highly diverse, and each dataset is unique in nature; therefore, obtaining a clear distinction between heterogeneous and homogeneous large online datasets is a very challenging task. The proposed online PC-KFA can reevaluate and change the criteria established in training the algorithm to correctly train the data. This allows us to efficiently differentiate polyps from false positives, in combination with a shape-based polyp detection method, and to improve the detection performance of CAD CTC.

The remainder of this chapter is organized as follows. Section 4.2 provides an introduction to kernels and a brief review of the existing kernel-based feature extraction methods of kernel principal component analysis (KPCA). In Section 4.3, we present PC-KFA for the detection of polyp candidates in the offline data. In Section 4.4, we extend this offline kernel method to accommodate online data, called online PC-KFA to select either homogeneous or heterogeneous on the basis of HBDA. Section 4.5 discusses the larger size of incoming data and its segmentation procedure. Section 4.6 presents the experimental results, and conclusions are drawn in Section 4.7.

4.2 KERNEL BASICS: A BRIEF REVIEW

We briefly review existing KPCA [24–27] in Section 4.2.1 and kernel selection in Section 4.2.2.

4.2.1 *Kernel Principal Component Analysis*

KPCA uses a Mercer kernel [28] to perform a linear principal component analysis of the transformed image. The eigenvalues λ_j and eigenvectors e_j are obtained by solving

$$\lambda_j e_j = S e_j = \sum_{i=1}^{n} \Phi(x_i)\Phi(x_i)^T e_j = \sum_{i=1}^{n} \langle e_j, \Phi(x_i)\rangle \Phi(x_i) \tag{4.1}$$

If K is an $n \times n$ Gram matrix, where the element $k_{ij} = \langle \Phi(x_i),\ \Phi(x_j)\rangle$ and $a_j = [a_{j1} a_{j2} \cdots a_{jn}]^T$ are the eigenvectors associated with eigenvalues $\lambda_{j,}$ then the dual eigenvalue problem equivalent to the problem can be expressed as follows:

$$\lambda_j a_j = K a_j$$

KPCA can now be summarized as follows:

1. Calculate the Gram matrix using kernels, which contains the inner products between pairs of image vectors.
2. Use $\lambda_j a_j = K a_j$ to get the coefficient vectors a_j for $j = 1, \ldots, n$
3. The projection of a test point $x \in R^d$ along the jth eigenvector is

$$\langle e_j, \Phi(x)\rangle = \sum_{i=1}^{n} a_{ji} \langle \Phi(x_i), \Phi(x)\rangle = \sum_{i=1}^{n} a_{ji} k(x, x_i) \tag{4.2}$$

The above implicitly contains an eigenvalue problem of rank n, so the computational complexity of KPCA is $O(n^3)$. In addition, each resulting eigenvector is represented as a linear combination of n terms.

The success of the KPCA depends heavily on the choice of kernel used for constructing the Gram matrix [29–31]. According to the No Free Lunch Theorem [32], there is no superior kernel function in general, and the performance of a kernel function rather depends on applications. This idea is discussed in the following section.

4.2.2 Kernel Selection

For efficient feature analysis, extraction of the salient features of polyps is essential because of the size and the three-dimensional nature of the polyp datasets. Moreover, the distribution of the image features of polyps is expected to be nonlinear. The problem is how to select such an ideal nonlinear operator. Some of the commonly used kernels are indexed as Kernel 1 to Kernel 4 [33–37]: (i) linear kernel, (ii) Gaussian radial basis function (RBF) kernel, (iii) Laplace RBF kernel, and (iv) Sigmoid kernel.

We adopt a data-dependent kernel [22] to capture the relationship among the data in this classification task using the composite form. This data-dependent composite kernel k_r for $r = 1, 2, 3, 4$ can be formulated as

$$k_r(x_i, x_j) = q_r(xi)q_r(xj)p_r(xi, xj) \tag{4.3}$$

where, $x \in R^d$, $p_r(xi, xj)$ is one kernel among four chosen kernels, and $q(.)$ is the factor function, with the following form for $r = 1, 2, 3, 4$; $q_r(x_i) = a_{r0} + \sum_{m=1}^{n} a_{rm} k_0(x_i,\ x_m)$, where $k_0(xi, xm) = \exp(-\|xi - xm\|^2/2\sigma^2)$ and a_{rm} are the combination coefficients.

Let the kernel matrices corresponding to $k(xi, xj)$, $p_r(xi, xj)$ be K_r, P_r. Therefore, we can express the data-dependent kernel K_r as

$$K_r = [q_r(xi)q_r(xj)p_r(xi, xj)]_{n\times n} \tag{4.4}$$

Defining Q_i as the diagonal matrix of elements $\{q_i(x_1), q_i(x_2), \dots, q_i(x_{xn})\}$ we can express as $K_r = Q_r P_r Q_r$.

The criterion for selecting the best kernel function is finding the kernel that produces largest eigenvalue [24–27]. The idea is that choosing the maximum eigenvector corresponding to the maximum eigenvalue that can result in the optimum solution. Once we derive the eigenvectors, that is, the combination coefficients of all four different kernels, we now proceed to construct $q_r (q_r = K_0 \alpha_r(n))$ and Q_r to find the corresponding Gram matrices of these kernels [38–41]. Once we have these Gram matrices ready, we can find the optimum kernels for the given dataset. To do this, we have to arrange the eigenvalues (that determined the combination coefficients) for all kernel functions in descending order.

Zheng [15] proposed a similar method to batch learning, where the input data was divided into a few groups of similar size, and KPCA was applied to each group. A set of eigenvectors was obtained for each group and the final set of features was obtained by applying KPCA to a subset of these eigenvectors. The application of the online concept to principal component analysis is mostly referred to incremental principal component analysis [11, 44–47] and has shown computational effectiveness in many image processing applications and pattern classification systems [42–43]. The effective application of online learning to nonlinear space is usually undertaken by using kernel-based methods [24–27].

4.3 KERNEL ADAPTATION ANALYSIS OF PC-KFA

A real CTC dataset suffers from nonlinear and diverse distributions among actual cancer datasets used for CAD [5], especially for the increased size of patients. By extending KPCA and kernel selection to address this data obstacle, we introduce an adaptive kernel algorithm called PC-KFA [21] with a composite kernel function that is defined as the weighted sum of the set of different optimized kernel functions for the training datasets as follows: $K_{comp}(\rho) = \sum_{i=1}^{p} \rho_i Q_i P_i Q_i$ where the value of the composite coefficient ρ_i is a scalar value, and p is the number of kernels we intend to combine in the training datasets. Through this approach, the relative contribution of all the kernels to the model can be varied over the input space when the new coming datasets change. Instead of using K_r as the old kernel matrix, we will be using $K_{comp}(\rho)$, which we call "kernel adaptation." According to [48], this composite kernel matrix $K_{comp}(\rho)$ satisfies Mercer's condition. Now the problem is how to determine this composite coefficient $\hat{\rho}(\hat{\rho} = [\rho_1, \rho_2, \dots, \rho_p])$ such that the classification performance is optimized. To this end, we used the concept of "kernel alignment" to determine the best $\hat{\rho}$ that gives us optimum performance to consider both offline and online datasets.

The "alignment" measure was introduced by Cristianini *et al.* [49] to measure the adaptability of a kernel to the target data and provide a practical objective for kernel optimization. The alignment measure is defined as a normalized Frobenius inner product between the kernel matrix and the target label matrix. The empirical alignment between kernel K_1 and kernel K_2 (label) with respect to the training set S is given as:

$$\text{Frob}(k_1, k_2) = \frac{\langle K_1, K_2 \rangle_F}{\|K_1\|_F \|K_2\|_F}$$

where K_i is the kernel matrix for the training set S using kernel function k_i, $\|K_i\|_F = \sqrt{\langle K_i, K_i \rangle_F}$, $\langle K_i, K_j \rangle_F$ is the Frobenius inner product between K_i and K_j. It has been shown that if a kernel chosen is well aligned with the other datasets, it does not change anything. If there exists a separation of the data with a low bound on the generalization error [49], it would be better to add one more kernel so that we can optimize the kernel alignment based on training both offline and online dataset to improve the generalization performance on the further test datasets. Let us consider the combination of kernel functions as: $k(\rho) = \sum_{i=1}^{p} \rho_i k_i$ where the kernels k_i, $i = 1, 2, \dots, p$ are known in advance. Our purpose is to tune ρ to maximize the empirical alignment between $k(\rho)$ and the target vector y. Hence,

$$\hat{\rho} = \arg\max_{\rho}(\text{Frob}(\rho, k_i, k_j)) = \arg\max_{\rho}\left(\frac{\left\langle \sum_i \rho_i K_i K_j \right\rangle}{\sqrt{\left\langle \sum_i \rho_i K_i \right\rangle, \left\langle \sum_j \rho_i K_j \right\rangle}}\right)$$

$$= \arg\max_{\rho}\left(\frac{\rho^T V_{ij}\rho}{\rho^T U_{ij}\rho}\right) \tag{4.5}$$

where $u_i = \sqrt{\langle K_i, yy^T \rangle}, v_{ij} = \sqrt{\langle K_i, K_j \rangle}, U_{ij} = u_i u_j, V_{ij} = v_i v_j, \rho = (\sqrt{\rho_1}, \sqrt{\rho_2}, \ldots,$ $\sqrt{\rho_p})$.

Let the generalized Raleigh coefficient be given as

$$J(\rho) = \frac{\rho^T V \rho}{\rho^T U \rho} \tag{4.6}$$

Therefore, we can obtain $\hat{\rho}$ by solving the generalized eigenvalue problem $V\rho = \delta U \rho$ where δ denotes the eigenvalues. Once we find this optimum composite coefficient $\hat{\rho}$, which will be the eigenvector corresponding to maximum eigenvalue δ, we can compute the composite data-dependent kernel matrix $K_{comp}(\rho)$. Because we have associated training of all offline and online data that make use of a data-dependent composite kernel, we can now proceed with the training of the data on the top. This extended framework has been developed, called PC-KFA [21].

Once we know the eigenvectors, that is, combination coefficients of the composite training data, we can compute q'_r and hence Q'_r to find out the four Gram matrices corresponding to the four different kernels. Now the first p kernels corresponding to the first p eigenvalues (arranged in the descending order) will be used in the construction of a composite kernel that will yield optimum classification accuracy. Before proceeding to the construction of the composite kernel, we have to find whether our new data is homogeneous or heterogeneous and update our Gram matrices accordingly.

4.4 HETEROGENEOUS VS. HOMOGENEOUS DATA FOR ONLINE PC-KFA

To make CT a viable option for screening larger patient populations over time, we propose HBDA to handle online mass data of clinical colon screening. The proposed HBDA may handle heterogeneous data acquired over a cumulative period of time to reevaluate the criteria. Online PC-KFA incorporates HBDA when the training properties of the data change dynamically. It is efficient to perform online learning if each pattern is presented in the limited storage space; thus PC-KFA using HBDA requires no additional memory for storing the patterns.

4.4.1 Updating the Gram Matrix of the Online Data

We use class separability as a measure to identify whether the data is heterogeneous or homogeneous. If the data is homogeneous, it improves the separability. Conversely, the heterogeneous data degrades the class separability ratio. Let us introduce a variable ξ, which is the ratio of the class separability of the composite online data and the offline data. It can be expressed as: $\xi = J'_*(\alpha'_r)/J_*(\alpha_r)$ where $J'_*(\alpha'_r)$ denotes the class separability yielded by the most dominant kernel for the composite data (i.e., new incoming data and the previous offline data). $J_*(\alpha_r)$ is the class separability yielded by the most dominant kernel for offline data. Thus, separability can be rewritten as: $\xi = \lambda'_*/\lambda_*$ where λ'_* corresponds to the most dominant eigenvalue of composite data (both offline and online), and λ_* is the most dominant eigenvalue of the four different kernels for the offline data. If ξ is less than a threshold value η (e.g., 1), then the incoming online data is heterogeneous; otherwise, it is homogeneous as described in 1) and 2) below, respectively. Figure 4.2 shows the relationship between separability and heterogeneous degree.

1) Homogeneous Online Data In the case of homogeneous data (via ξ), we do not update the Gram matrix. Instead, we discard all the data that is homogeneous. Hence, the updated Gram matrix can be given as $K_r^{n'} = Q_r P_r Q_r$.

2) Heterogeneous Online Data If the new sequence of incoming data is heterogeneous, we update our feature space. However, this update can be either incremental or nonincremental. We propose a criterion called "heterogeneous degree" $\mathrm{HD} = (\overline{\alpha}'_* - \overline{\alpha}_*)^{\leftrightarrow}\lambda'_*/\lambda_*$ to determine if the data is highly heterogeneous or less heterogeneous. The class separability of the most dominant kernel for the new data is directly dependent on both the maximum combination coefficient α'_* (this is the maximum combination coefficient of four different kernels) as well as the maximum eigenvalue λ'_*. Let us denote $\overline{\alpha_*}$ as the mean of combination coefficients, and $\overline{\alpha'_*}$ is the most dominant kernel among the four kernels available.

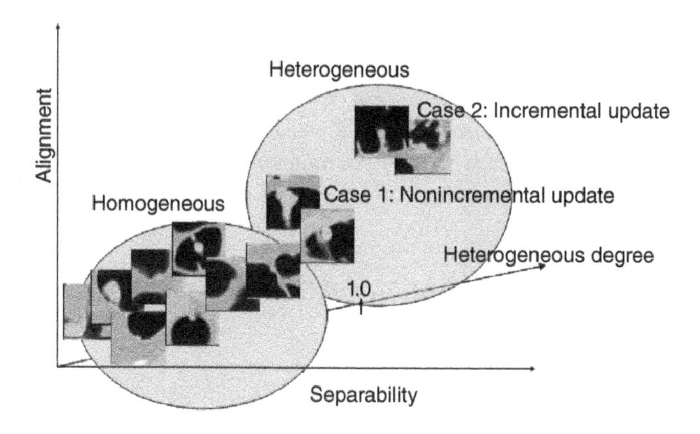

Figure 4.2 Criteria concept for homogeneous and heterogeneous online data.

If the heterogeneous degree HD is less than 1 (Case1), then the update is nonincremental. Hence, the dimensions of the Gram matrix remain constant. The input matrix P' and Q' are updated by replacing the row and column corresponding to the index.

If the heterogeneous degree is greater than 1 (Case 2), then this means the difference between the previous eigenvectors and the new eigenvectors is large. So it is very important for us to retain this new data, which is highly heterogeneous for efficient classification. We simply retain it and hence the size of the Gram matrix is incremented by the size of the new data. In this case, input matrix P' and Q' are the same as in Section 4.2.2, as we have already calculated the new combination coefficient α'_r.

Obtaining a higher dimensional feature space of the training data with greater classification power depends on how effectively we are updating the Gram matrix. Because construction of the Gram matrix is the crucial step for performing feature analysis, any redundant data in the Gram matrix should be preserved for further training of the algorithm.

Once we have our kernel Gram matrices for Case 1 and 2, we can now determine the composite kernel that will give us the optimum classification accuracy when the existing offline data is incorporated with new online data.

4.4.2 *Composite Kernel for Online Data*

As we have computed the Gram matrices for p, most dominant kernels for the composite data (offline plus online data), we now combine them to yield the best classification performance. As defined in Section 4.2.2, we can write the composite kernel for the new composite data:

$$K^{n'}_{comp}(\rho) = \sum_{i=1}^{p} \rho'_i Q_i^{n'} P_i^{n'} Q_i^{n'} \tag{4.7}$$

We use the same alignment factor method to determine the optimum composite coefficients that will yield the best composite kernel. Therefore, the new alignment factor can be recalculated as Equation 4.5. We can determine the composite coefficient ρ' that will maximize the alignment factor as Equation 4.6 of the new generalized Raleigh coefficient (Fig. 4.3).

We can obtain the value of $\widehat{\rho}'$ by solving the generalized eigenvalue problem $U^{n'}\rho' = \delta' V^{n'}\rho'$ where δ' denotes the eigenvalues. We choose a composite coefficient vector that is associated with the maximum eigenvalue. Once we determine the best composite coefficients, we can compute the composite kernel.

4.5 LONG-TERM SEQUENTIAL TRAJECTORIES WITH SELF-MONITORING

The CTC data acquired over a long period of time can sometimes be highly diverse; that is, the portion of online datasets is dominant for the CAD system. There are no particular methods yet to differentiate between heterogeneous and homogeneous big

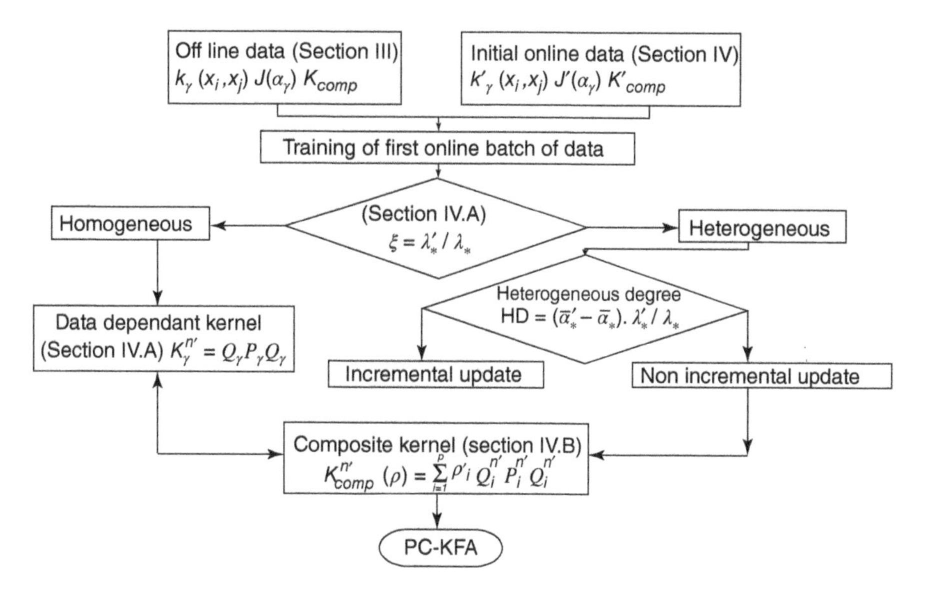

Figure 4.3 Overall flow of online data training using HBDA with PC-KFA.

online CTC data including online virtual colonography. To facilitate online training of long-term sequential trajectories, we need a new method that requires retention of only a portion of the training data due to the size of long sequential datasets. The challenge herein is in determining what data to retain. Hence, in our study, we present a self-monitoring way of classifying the incoming long sequential data as either heterogeneous or homogeneous and automatically updating the feature space over time accordingly.

4.5.1 Reevaluation of Large Online Data

One way is to track the online data using the "alignment" measure for the kernel adaptability of long sequential online datasets to the other existing online datasets. We extend the alignment measure of the normalized Frobenius inner product with the time sequential index. If there is no compromise when we train the algorithm without further modifications, then there is no need to break down the incoming sequence of data into small windows or change any parameters in the previous setting. Let us denote the time index ${}^{t+1}(K^{n'}_{comp})_t$ as the update of the Gram matrix from time t to $t+1$, and let ${}^{t}(K^{n'}_{comp})_0$ be the updated Gram matrix from time '0' (i.e., from the beginning till one last step) to time t. The alignment factor can be respectively given as ${}^{t+1}(\mathrm{Frob}({}^{t+1}(K^{n'}_{comp})_t, {}^{t}(K^{n'}_{comp})_0))_t$ and ${}^{t}(\mathrm{Frob}({}^{t+1}(K^{n'}_{comp})_t, {}^{t}(K^{n'}_{comp})_0))_0$. Owing to the simplicity, we may drop Gram matrix notation ${}^{t+1}(K^{n'}_{comp})_t$ in the Frob term. The proposed criterion is to track the erroneous training of the data by comparing these alignment factors as follows:

$${}^{t+1}(\mathrm{Frob})_t \;<^{t}\; (\mathrm{Frob})_0$$

If the alignment factor deteriorates, the margin parameter is set to a different value and the experiment must be run again for the optimal use of PC-KFA. A significant variation in the alignment factor shows that the incoming online data is too large to correctly train it as either heterogeneous or homogeneous in the next stage. In this case, we divide and keep dividing the data into several small chunks of data until we find a suitable window size of online data that would be appropriate for training and yield the best results.

4.5.2 Validation of Decomposing Online Data into Small Chunks

Previously, we divided the new online data into equal sized subdatasets, which we called $d1,\ d2,\ d3, \ldots$ where the size of each new subdataset was equal to 1. We extend this restriction for the long sequential datasets as heterogeneous or homogenous data by checking the class separability of each dataset to find the appropriate window size for the incoming data and determine whether it is homogeneous or heterogeneous data using the updated criteria. Figure 4.4 illustrates the segmentation of the incoming online data into small equal subsets if the incoming online data is heterogeneous. If the data is homogeneous, most of the data is redundant, and hence we simply discard this homogeneous data, and we do not update any information from the new subset on to the previous Gram matrix. Hence, the kernel Gram matrix at time $t+1$ is same as the previous one at time t. On the contrary, if the data is heterogeneous, we update the Gram matrix either incrementally or nonincrementally depending on the level of heterogeneous degree of the online data.

The new criterion for the decomposition of online data into small chunks is time sequential class separability of the composite data (data up to time t and $d1$ data) as

$$^{t+1}(\xi)_t = \frac{^{t+1}(\lambda'_*)_t}{^{t}(\lambda_*)_0} \tag{4.8}$$

where $^{(t+1)}(\lambda'_*)_t$ is the largest eigenvalue of the data (of the dominant kernel) received from time t to $t+1$, and $^{t}(\lambda_*)_0$ is the largest eigenvalue of the dominant kernel

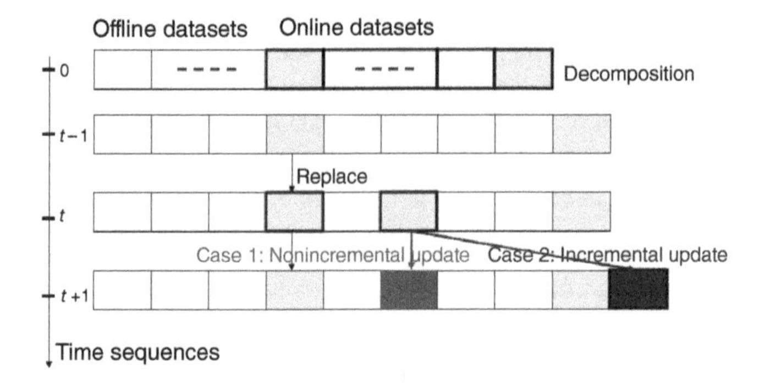

Figure 4.4 Online decomposition for heterogeneous sequences.

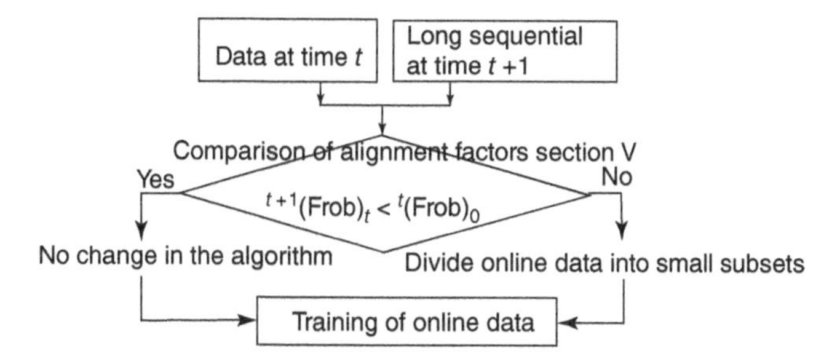

Figure 4.5 Training of online datasets acquired over long-term sequences.

calculated from time *0* to time t. If ${}^{(t+1)}(\xi)_t$ is greater than a new threshold value η', then the incoming data is heterogeneous; otherwise, it is homogeneous.

If the deviation of alignment factor is huge, then the step size is further reduced and the algorithm is computed again. This process should be repeated until we find an appropriate window size of the incoming data that would allow us for the proper classification of the homogeneous and heterogeneous data and result in less-error training. After the training of the Gram matrix is finished to incorporate the dynamic features of the new online data, the PC-KFA algorithm is applied to the kernel Gram matrix. This entire process is summarized in the flow chart in Fig. 4.5.

4.6 EXPERIMENTAL RESULTS

The experimental results consist of Sections 4.6.1–4.6.6.

4.6.1 Cancer Datasets

The proposed online PC-KFA together with HBDA was evaluated using CTC image datasets of colonic polyps composed of true positive (TP) and false positive (FP) polyps detected by our CAD system [5]. We obtained studies of 146 patients who had undergone a colon-cleansing regimen, such as optical colonoscopy. These CTC cases were acquired by eight different models of single- and multidetector helical scanners by use of $1.25 - 5.0$ mm collimations, pitch of 1–2, reconstruction interval of $1.25 - 5.0$ mm, and tube current of $50 - 200$ mA. The patients were followed up by conventional optical colonoscopy, which served as the gold standard for the presence of polyps. Among 464 CTC cases, 59 cases were abnormal (having at least one polyp ≥ 6 mm in size) and 405 cases were normal (having no colonic lesson). Each patient was scanned in both supine and prone positions; thus each CTC case consisted of two reconstructed CTC volumes, resulting in 928 CTC volumes with effective voxel size of 0.5 mm/voxel. The volumes of interest (VOIs) representing each polyp candidate have been calculated as follows. The CAD scheme provided a segmented region for each candidate. The center of the VOI was placed at the center of mass of the region. The size of the VOI was chosen so that the entire region was covered. The resampling

Table 4.1 Arrangement of offline datasets

	Number of vectors in training set			Number of vectors in test set		
Data	TP	FP	Total	TP	FP	Total
Dataset1	21	69	90	8	32	40
Dataset2	38	360	398	16	300	316
Dataset3	10	500	510	6	425	431
Dataset4	7	1050	1057	4	1200	1204

has been carried out using VOIs with dimensions 12 × 12 × 12 voxels to build the Dataset1 that consists of 29 true polyps and 101 false polyps. For the rest of the datasets, the VOI was resampled to 16 × 16 × 16 voxels. The VOIs so computed comprise the Dataset2 that consists of 54 TPs and 660 FPs, Dataset3 that consists of 16 TPs and 925 FPs, and the Dataset4 that consists of 11 TPs and 2250 FPs.

Table 4.1 shows the arrangement of offline training and test sets for the colon cancer datasets. Instead of using cross-validation, we randomly divide the entire data into training and testing, as they are highly imbalanced: most datasets are FP, and very few are TP.

4.6.2 Selection of Optimum Kernel and Composite Kernel for Offline Data

We used the method proposed in Section 4.3 to create four different data-dependent kernels and select the kernel that best fits the offline CTC data listed in Table 4.1. We determined the optimum kernel depending on the eigenvalue that produces maximum separability. Table 4.2 indicates the eigenvalues λ and hyperparameters of four kernels for each dataset.

The two kernels with the two maximum eigenvalues are highlighted for each offline dataset in Table 4.2. In Dataset1, for example, we combined RBF and Laplace to form the composite kernel. We have observed that each database had different combinations for the composite kernels. The composite coefficients of the two most

Table 4.2 Eigenvalues of four kernels for offline datasets

Data	Linear	Poly	RBF	Laplace	Combination
Dataset1	$\lambda = 10.66$	$\lambda = 10.25$	$\boldsymbol{\lambda = 14.13}$	$\boldsymbol{\lambda = 12.41}$	*RBF 0.98*
		$d = 1.2$, *Offset* $= 2$	$\sigma = 4.12$	$\sigma = 0.9$	*Laplace 0.14*
Dataset2	$\lambda = 102.08$	$\boldsymbol{\lambda = 105.91}$	$\boldsymbol{\lambda = 116.64}$	$\lambda = 80.57$	*RBF 0.72*
		$d = 1$, *Offset* $= 4$	$\sigma = 5.29$	$\sigma = 3.5$	*Linear 0.25*
Dataset3	$\boldsymbol{\lambda = 57.65}$	$\lambda = 51.35$	$\boldsymbol{\lambda = 74.55}$	$\lambda = 30.23$	*RBF 0.98*
		$d = 1.4$, *Offset* $= 1$	$\sigma = 5.65$	$\sigma = 1.0$	*Linear 0.23*
Dataset4	$\lambda = 72.41$	$\boldsymbol{\lambda = 83.53}$	$\boldsymbol{\lambda = 124.13}$	$\lambda = 56.35$	*RBF 0.91*
		$d = 0.8$, Offset $= 2$	$\sigma = 4$	$\sigma = 2.5$	*Poly 0.18*

dominant kernels are listed in the last column as described in Section 4.3. For the four CTC datasets we have, the most dominant kernel was the RBF kernel, whereas the second most dominant kernel kept varying.

The classification accuracy and mean square reconstruction error for the offline datasets are documented in Table 4.3. The mean square reconstruction error in the case of offline data using PC-KFA algorithm was calculated as $Err_i = \|\Phi_i - \Phi_i'\|^2$. The classification accuracy was calculated as (TP + TN)/(TP + TN + FN + FP) where TN stands for "true negative," TP stands for "true positive," FP stands for "false positive," and FN stands for "false negative." Figure 4.6 is receiver operating characteristic (ROC) curve for false positive rate (FPR) versus true positive rate (TPR) with the calculated region of ROC, to compare the four datasets on the basis of the area under curve (AUC).

From Fig. 4.6 and Table 4.3, we obtained good classification accuracy and ROC performance by using PC-KFA over all the offline datasets. All four CTC image datasets of colonic polyps were well detected using PC-KFA [21], and the largest dataset, Dataset4, achieved 99.67% classification accuracy. In the following sections,

Table 4.3 Classification accuracy and mean square error of offline data

Data	Mean square reconstruction error of offline data with composite kernel, %	Classification accuracy of offline data with composite kernel, %
Dataset1	1.0	90
Dataset2	9.64	94.62
Dataset3	6.25	98.61
Dataset4	14.03	99.67

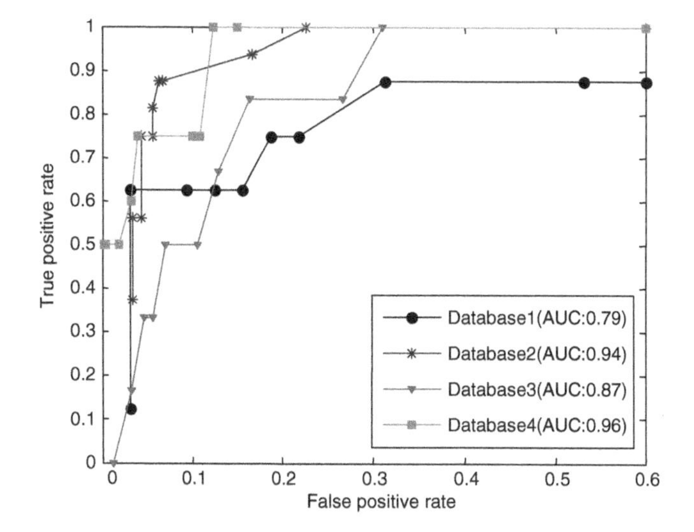

Figure 4.6 The ROC performance of offline data using PC-KFA.

we further evaluate the results of this method extended to online HBDA, which included heterogeneous big datasets of CTC image.

4.6.3 Selection of Optimum Kernel and Composite Kernel for the new Online Sequences

We followed Section 4.4 to tune the selection of appropriate kernels when new online data became available. Table 4.4 shows such a new online data stream called "Online Sequence ##." Each of the data sequences originates from the same original cumulated database. They differed in the fraction of the database, which was used for initial training. Online Sequence #1, Online Sequence #2, Online Sequence #3, and Online Sequence #4 were defined in Section 4.5.2, and generated in the amount of 0.5%, 5%, 10%, and 20% of the original cumulated database used for initially labeled training. The term "Online Sequence" was also explained in Fig. 4.7. This figure shows the relationships between Datasets and Online Sequences, corresponding to Table 4.4.

After we tentatively formed the input matrices for four different kernels, we used Equation 4.6 to find the dominant kernels for the new data and the previous offline data. These results are summarized in Table 4.5.

Table 4.5 shows the dominant kernels with the bolded λ as the largest eigenvalue. As seen in the online data sequences, the eigenvalues calculated were gradually

Table 4.4 Online sequences generated from datasets

Data	Online Sequence #1	Online Sequence #2	Online Sequence #3	Online Sequence #4
Dataset1	3 TP, 12 FP	3 TP, 12 FP	3 TP, 12 FP	3 TP, 12 FP
Dataset2	7 TP, 85 FP	7 TP, 85 FP	7 TP, 85 FP	7 TP, 85 FP
Dataset3	2 TP, 87 FP	2 TP, 87 FP	2 TP, 87 FP	2 TP, 87 FP
Dataset4	2 TP, 126 FP	2 TP, 126 FP	2 TP, 126 FP	2 TP, 126 FP

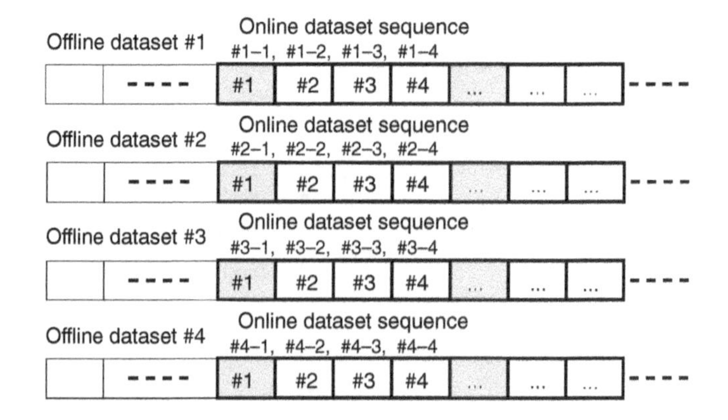

Figure 4.7 The online dataset sequences of Table 4.4.

Table 4.5 Eigenvalues of four different kernels (linear, poly, RBF, Laplace) with λ for online data sequences

Data		Online Sequence #1	Online Sequence #2	Online Sequence #3	Online Sequence #4
Dataset1	Linear	*9.99***	*9.89*	*10.16*	*9.73*
	Poly	*9.31***	*8.96*	*9.52*	*10.31*
	RBF	***11.34*****	***12.47***	***18.09***	***24.69***
	Laplace	*8.41***	*6.7*	*12.85*	*13.27*
Dataset2	Linear	*129.44*	*84.97***	*95.63***	*109.02*
	Poly	*106.44*	*85.01***	*96.27***	*109.61*
	RBF	***134.59***	***101.21*****	***122.81*****	***133.40***
	Laplace	*90.76*	*38.49***	*84.12***	*115.78*
Dataset3	Linear	*75.13*	*73.88*	*62.59***	*72.45*
	Poly	*36.38*	69.25	*66.48***	*83.39*
	RBF	***80.16***	***86.41***	***73.66*****	***109.29***
	Laplace	*44.66*	*13.87*	*49.46***	*50.25*
Dataset4	Linear	*82.72*	*52.99***	*69.43*	*98.64*
	Poly	*92.33*	*75.83***	*79.85*	*112.39*
	RBF	***132.47***	***103.25*****	***144.59***	***157.27***
	Laplace	*38.67*	*43.26***	*32.59*	*62.85*

"**" Indicates the eigenvalues after updating the matrix according to the method presented in Section 4.4.

shifting, but the dominant kernels choice remained the same. Therefore, PC-KFA consistently maintained the choice of dominant (and second dominant) kernels to update data-dependent kernel matrices for the computation of the composite kernel matrix. Under the screening of large patient populations, the detection of colonic polyps needs further evaluation by associating the existing data with newly acquired online sequences. Thus, we will evaluate more HBDA characteristics of long-term online data sequences in the next subsections.

4.6.4 Classification of Heterogeneous versus Homogeneous Data

After determining the dominant two kernels, the next step was to update these data-dependent kernel matrices for the computation of the composite kernel matrix. We imported the criterion 4.8 in Section 4.5 to classify this new data to the two categories, "homogeneous" and "heterogeneous."

Table 4.6 shows that the majority of online data sequences was homogeneous, in which the Gram matrix in this sequence was not updated to save computation time. If the database was homogenous, we set the data-dependent kernels as "No update." However, the HD determines Case 1 as "Nonincremental Update," or Case 2 as "Incremental Update," for all the heterogeneous data sequences in Section 4.5.1. The results for all the online data sequences are given in the following Table 4.7.

Once we have the data-dependent kernel Gram matrices for four different kernels, we can now proceed to calculate the data-dependent composite kernels for the new

Table 4.6 Homegeneous and heterogeneous categories of online sequences

Data	Online Sequence #1	Online Sequence #2	Online Sequence #3	Online Sequence #4
Dataset1	**Heterogeneous**	Homogeneous	Homogeneous	Homogeneous
Dataset2	Homogeneous	**Heterogeneous**	**Heterogeneous**	Homogeneous
Dataset3	Homogeneous	Homogeneous	**Heterogeneous**	Homogeneous
Dataset4	Homogeneous	**Heterogeneous**	Homogeneous	Homogeneous

The bold fonts indicate the Hererogeneous cases.

Table 4.7 Update of different sets of Online Sequences with Gram matrix size

Data	Online Sequence #1	Online Sequence #2	Online Sequence #3	Online Sequence #4
Dataset1	Nonincremental 85×85	No update 5×85	No update 5×85	No update; 85×85
Dataset2	No update 398	Nonincremental 398	Nonincremental 398	No update 398
Dataset3	No update 510	No update 510	Nonincremental 510	No update 510
Dataset4	No update 1057	Nonincremental 1057	No update 1057	No update 1057

online data sequences on the basis of the method presented in Section 4.5.2 as shown in Table 4.8.

Table 4.8 shows the composite kernel coefficients for the two most dominant kernels. The classification accuracy and mean square reconstruction error of the online data sequences were calculated in Section 4.5. Table 4.8 shows that the classification accuracy for online sequence had an average of 95.12% with variance 2.03%. This was comparable to the offline data performance, which yielded a classification accuracy of 95.72% with variance 4.39%.

4.6.5 Online Learning Evaluation of Long-term Sequence

In this subsection, we evaluate the performance of online HBDA with PC-KFA using long-term data sequences that are much larger than the online sequences analyzed in previous subsections. Figure 4.8 shows data distribution of long-term sequential trajectories as the evaluation of long-term online learning.

For the long-term sequential trajectory, the experimental dataset was composed of all the available images accumulated. A preset percentage of images was selected randomly from the dataset and used for the offline training stage using PC-KFA. The rest of the data were sequentially fed to the online HBDA algorithm for evaluation

Table 4.8 Classification accuracy and MSE with composite kernels of online data sequences

Data	Online Sequence #1	Online Sequence #2	Online Sequence #3	Online Sequence #4
Dataset1	MSE: 0.43 Accuracy: 92.5 RBF: $\rho_1 = 0.99$ Linear: $\rho_2 = 0.17$	sab	sab	sab
Dataset2	sab	MSE: 15.06 Accuracy: 93.99 RBF: $\rho_1 = 0.93$ Linear: $\rho_2 = 0.16$	MSE: 14.50 Accuracy: 94.94 RBF: $\rho_1 = 0.99$ Linear: $\rho_2 = 0.14$	sab
Dataset3	sab	sab	MSE: 7.68 Accuracy: 96.52 RBF: $\rho_1 = 0.73$ Poly: $\rho_2 = 0.32$	sab
Dataset4	sab	MSE: 15.63 Accuracy: 97.65 RBF: $\rho_1 = 0.90$ Poly: $\rho_2 = 0.28$	sab	sab

Abbreviation: sab, same as before.

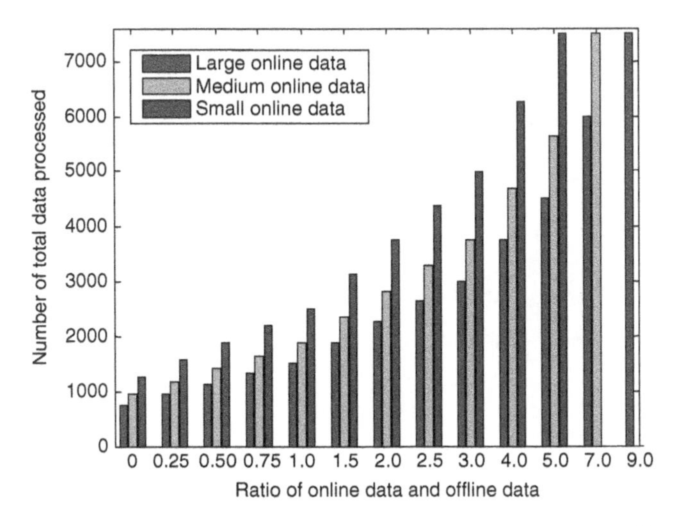

Figure 4.8 Long-term data sequence used to evaluate online learning. The horizontal axis denotes the ratio of number of online data to number of offline data. The vertical axis denotes the number of total data processed using online HBDA with PC-KFA. The three long sequences are labeled as "Large Online Data," "Medium Online Data," and "Small Online Data," which were used corresponding to offline training dataset size of 750, 937, and 1250, respectively.

of the long-term sequential trajectory. The different ratios of training data between online and offline were prepared as in Fig. 4.8 from 0 to 9; for example, 5 means that online data was five times bigger than offline datasets. Note that we had three different cases of offline data, so "Large Online Data" was large online data used compared to relatively small offline data. We analyze how these ratios affect the online learning performance for AF, AUC, and classification accuracy.

AF was calculated using several long-term online sequences. If no increase, we divided this online data sequence of data into small subsets using the threshold value with online HBDA with PC-KFA. These results are shown in Fig. 4.9.

Figure 4.9 shows that AF was increased when more data was used to train the online HBDA with PC-KFA. As the ratio of the online to offline size increased, the AF increased for all three, small, medium, and large online data. Albeit more diverse data were fed in the form of online sequences from larger patients, the proposed online HBDA with PC-KFA adapted itself, as shown in the increase of AF. Figure 4.9 shows that AF consistently increased for long-term sequences of "large/medium/small" online data using online HBDA with PC-KFA even if heterogeneous characteristics was affected as in the cases of Tables 4.7 and 4.8.

The online HBDA with PC-KFA was evaluated on the basis of AUC for the online long-term sequences. Figure 4.10 shows that online PC-KFA exhibited an AUC performance similar to Fig. 4.6 for classification in handling the online data in all three different long-term sequences. The ability to track changes using long-term online sequences was also verified by the results shown in the following Fig. 4.11.

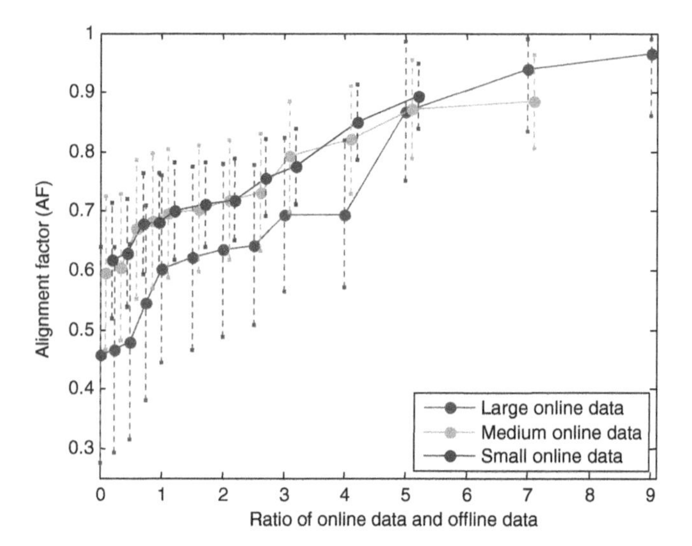

Figure 4.9 Alignment factors for long-term sequences. The horizontal axis denotes the progress of the online HBDA (PC-KFA) with time (ratio of online and offline training). The solid lines denoted the mean value of the observed AF value, while the dashed lines show the range of observed AF.

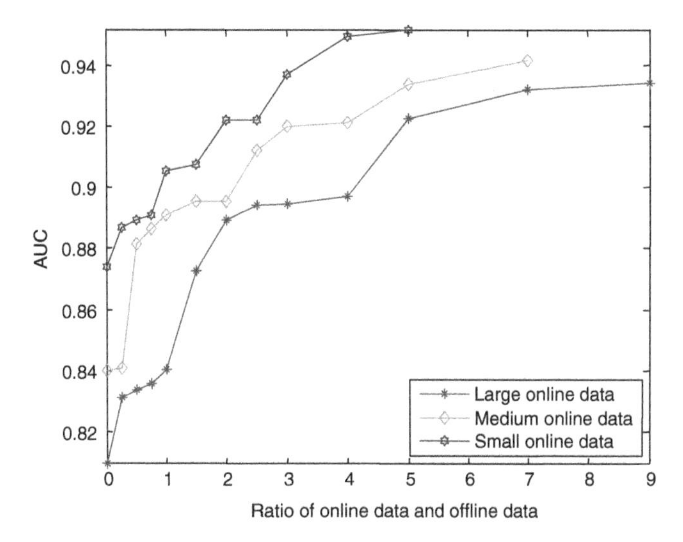

Figure 4.10 AUC of Receiver operating Characteristics for three long-term sequences. The horizontal axis denotes the progress of the online HBDA (PC-KFA) with time (ratio of online-to-offline training).

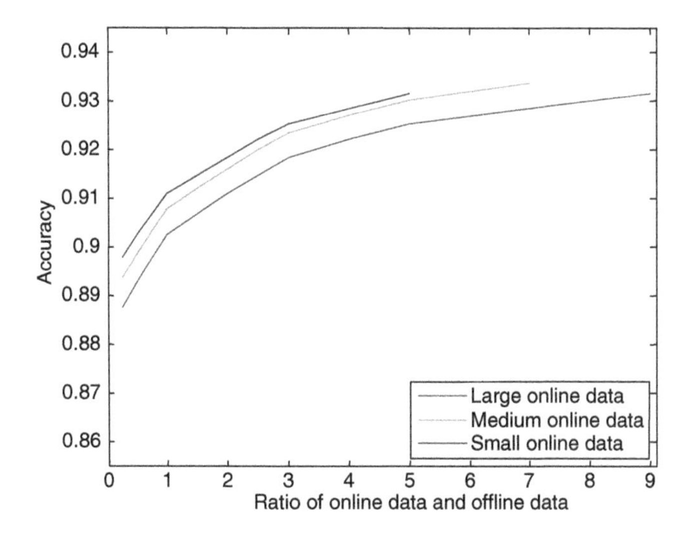

Figure 4.11 Accuracy versus ratio of online data to offline data.

Figure 4.11 shows that the proposed online HBDA with PC-KFA, handling very large online data over long-term sequences, performed with similar classification accuracy as in Table 4.3 of Section 4.6.2. After a finite number of sequences, the classification performance of the online data sets was approaching that of offline training data. This indicates that training of the subsequent online data sequence was advantageous over statistical offline learning. Due to the reduced size of kernel space, although the data size increased, the proposed online HBDA with PC-KFA achieved a consistent performance of classification accuracy.

4.6.6 Evaluation of Computational Time

In this subsection, we analyze the time required for processing of HBDA using online PC-KFA. All the experiments were implemented in MATLAB® R2010a using the Statistical Pattern Recognition Toolbox for the Gram matrix calculation and kernel projection. For processing the large volume of data, an Intel® core i7 of 3.40 GHz clock speed was used along with a workstation containing 16 GB system memory. Run time was determined using the *cputime* command. Because we do not consider all available online data after the initial offline training, the proposed HBDA is expected to yield significant savings regarding computational time.

Figure 4.12 shows total computational time (sum of times for both offline and online) for three long-term online sequences with different sizes. The mean training time was computed as the total training time divided by the number of samples processed.

Table 4.9 shows the individual means of offline and online training from Fig. 4.2. Table 4.9 clearly demonstrates that the computation time for online training was much smaller than that for offline training; in average, a spectacular 93% reduction of computational time per sample was achieved. Figure 4.12 also shows that the calculation of online training was computationally efficient as the ratio of online to offline training increased. Therefore, HBDA using online PC-KFA was better-suited to handle long-term sequences in a real-time manner. To make CTC a more acceptable technique for larger screening datasets, our computational speed for larger CTC database was promising.

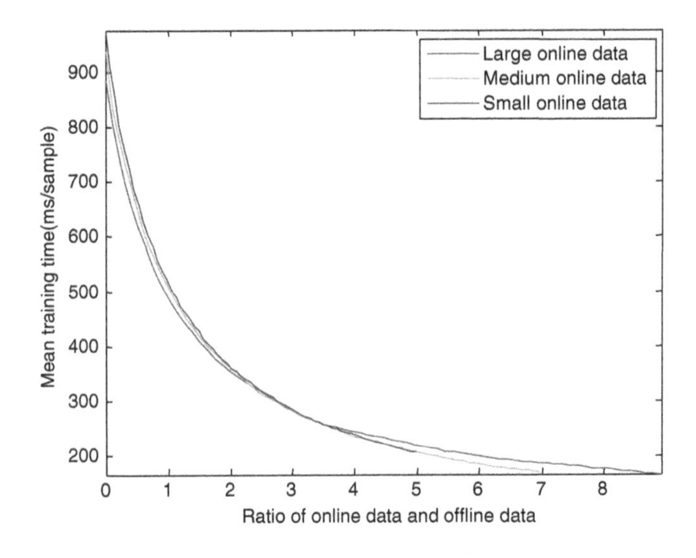

Figure 4.12 Computational time for online HBDA with offline PC-KFA.

Table 4.9 Online data sequence computation time

Online data size	Mean offline training time, ms/sample	Mean online training time, ms/sample
Large	884.34	85.95
Medium	936.65	59.99
Small	976.16	52.51

4.7 CONCLUSIONS

This chapter introduced an emerging and promising alternative to rigid colonoscopy for the detection of precancerous lesions in the colon. We proposed HBDA to handle the long-term heterogeneous trajectories of online data using PC-KFA. We applied data-dependent composite kernels to colon cancer datasets by maximizing a measure of class separability in the empirical feature space of online datasets. The composite combination vector (i.e., weight vector) for the most dominant kernels was determined by maximizing the alignment measure. We used the properties of heterogeneous degree to dynamically adjust to changes in the heterogeneous data during the online training. This online method was a faster and more efficient feature extraction algorithm to act as an efficient means for the detection of polyps on CTC images. The polyp classification experiment showed that online PC-KFA gave comparable classification performance of online data trained with that of the initial offline training data. HBDA applied to long-term data sequences in a model-based CAD scheme may yield high detection performance of polyps. Such an online CAD framework has a potential of making CT a viable option for screening large patient populations, resulting in early detection of colon cancer and leading to reduced mortality due to colon cancer. The proposed online HBDA with PC-KFA will reduce radiologist workload and improve diagnostic accuracy of larger patients by increasing true positive CTC cases.

REFERENCES

1. S. Winawer, R. Fletcher, D. Rex, J. Bond, R. Burt, J. Ferrucci, T. Ganiats, T. Levin, S. Woolf, D. Johnson, L. Kirk, S. Litin, and C. Simmang, "Colorectal cancer screening and surveillance: Clinical guidelines and rationale-Update based on new evidence," *Gastroenterology*, vol. 124, pp. 544–560, 2003; *Systems*, vol. 36, no. 1, pp. 266–278, 2000.
2. K. D. Bodily, J. G. Fletcher, T. Engelby, M. Percival, J. A. Christensen, B. Young, A. J. Krych, D. C. Vander Kooi, D. Rodysill, J. L. Fidler, and C. D. Johnson, "Nonradiologists as second readers for intraluminal findings at CT colonography," *Acad. Radiol.*, vol. 12, pp. 67–73, 2005.
3. J. G. Fletcher, F. Booya, C. D. Johnson, and D. Ahlquist, "CT colonography: Unraveling the twists and turns," *Curr. Opin. Gastroenterol.*, vol. 21, pp. 90–98, 2005.
4. D. Hock et al, "Virtual dissection CT colonography: Evaluation of learning curves and reading times with and without computer-aided detection," *Radiology*, vol. 248, no. 3, pp. 860–868, 2008.

5. H. Yoshida and J. Näppi, "Three-dimensional computer-aided diagnosis scheme for detection of colonic polyps," *IEEE Trans. Med. Imaging*, vol. 20, pp. 1261–1274, 2001.
6. H. Yoshida, Y. Masutani, P. MacEneaney, D. T. Rubin, and A. H. Dachman, "Computerized detection of colonic polyps at CT colonography on the basis of volumetric features: Pilot study," *Radiology*, vol. 222, pp. 327–36, 2002.
7. J. Näppi and H. Yoshida, "Fully automated three-dimensional detection of polyps in fecal-tagging CT colonography," *Acad. Radiol.*, vol. 14, pp. 287–300, 2007.
8. H. Yoshida and J. Näppi, "CAD in CT colonography without and with oral contrast agents: Progress and challenges," *Comput. Med. Imaging Graph.*, vol. 31, pp. 267–84, 2007.
9. T. A. Chowdhury, P. F. Whelan, and O. Ghita, "A fully automatic CAD-CTC system based on curvature analysis for standard and low-dose CT data," *IEEE Trans. Biomed. Eng.*, vol. 55, no. 3, pp. 888–901, 2008.
10. J. W. Suh and C. L. Wyatt, "Registration under topological change for CT colonography," *IEEE Trans. Biomed. Eng.*, vol. 58, no. 5, pp. 1403–1411, 2011.
11. L. Lu, D. Zhang, L. Li, et al., "Fully automated colon segmentation for the computation of complete colon centerline in virtual colonoscopy," *IEEE Trans. Biomed. Eng.*, vol. 59, no. 4, pp. 996–1004, 2012.
12. P. Lefere, S. Gryspeerdt, and A. L. Baert, *Virtual colonoscopy: A practical guide*. Springer-Verlag New York, LLC, New York, New York, 2009.
13. H. Cevikalp, M. Neamtu, and A. Barkana, "The kernel common vector method: A novel non-linear subspace classifier for pattern recognition," *IEEE Trans. Syst. Man Cyb. B*, vol. 37, no. 4, pp. 937–951, 2007.
14. B. J. Kim, I. K. Kim, and K. B. Kim, "Feature extraction and classification system for non-linear and online data," in *Advances in Knowledge Discovery and Data mining Proc.*, vol. 3056, pp. 171–180, 2004.
15. W. Zheng, C. Zou, and L. Zhao, "An improved algorithm for kernel principal component analysis," *Neural Process. Lett.*, vol. 22, no. 1, pp. 49–56, 2005.
16. M. Awad, Y. Motai, J. Nappi, and H. Yoshida, "A clinical decision support framework for incremental polyps classification in virtual colonoscopy," Special Issue on Machine Learning for Medical Imaging, *Algorithms*, vol. 3, pp. 1–20, 2010.
17. L. Winter, Y. Motai, and A. Docef, "Computer-aided detection of polyps in CT colonography: On-line versus off-line accelerated kernel feature analysis," Special Issue on Processing and Analysis of High-Dimensional Masses of Image and Signal Data, *Signal Process.*, vol. 90, pp. 2456–2467, 2010.
18. W. Cai, J.-G. Lee, E. M. Zalis, and H. Yoshida, "Mosaic decomposition: An electronic cleansing method for inhomogeneously tagged regions in noncathartic CT colonography," *IEEE Trans. Med. Imaging*, vol. 30, no. 3, pp. 559–574, 2011.
19. X. H. Xiong and M. N. S. Swamy, "Optimizing the kernel in the empirical feature space," *IEEE Trans. Neural Netw.*, vol. 16, no. 2, pp. 460–474, 2005.
20. J. Ye, S. Ji, and J. Chen, "Multi-class discriminant kernel learning via convex programming," *J. Mach. Learn. Res.*, vol. 9, pp. 719–758, 1999.
21. Y. Motai and H. Yoshida, "Principal composite kernel feature analysis: Data-dependent kernel approach," *IEEE Trans. Knowl. Data Eng.*, vol. 25, no. 8, pp. 1863–1875, 2013.
22. X. Jiang, R. Snapp, Y. Motai, and X. Zhu, "Accelerated kernel feature analysis," in *Proc. IEEE Computer Society Conference on Computer Vision and Pattern Recognition*, pp. 109–116, 2006.

23. V. N. Vapnik, *The nature of statistical learning theory*, 2nd ed. New York: Springer, 2000.
24. B. J. Kim and I. K. Kim, "Incremental non-linear PCA for classification," in *Proc. Knowledge Discovery in Databases (PKDD 2004)*, Pisa, Italy, September, 2004, vol. 3202, pp. 291–300, 2004.
25. B. J. Kim, J. Y. Shim, C. H. Hwang, I. K. Kim, and J. H. Song, "Incremental feature extraction based on empirical kernel map," *Found. Intell. Syst.*, vol. 2871, pp. 440–444, 2003.
26. L. Hoegaerts, L. De Lathauwer, I. Goethals, J. A. K. Suykens, J. Vandewalle, and B. De Moor, "Efficiently updating and tracking the dominant kernel principal components," *Neural Netw.*, vol. 20, no. 2, pp. 220–229, 2007.
27. T. J. Chin and D. Suter. "Incremental kernel principal component analysis," *IEEE Trans. Image Process.*, vol. 16, no. 6, pp. 1662–1674, 2007.
28. D. S. Paik, C. F. Beaulieu, G. D. Rubin, B. Acar, R. B. Jeffrey, Jr., J. Yee, J. Dey, and S. Napel, "Surface normal overlap: a computer-aided detection algorithm with application to colonic polyps and lung nodules in helical CT," *IEEE Trans. Med. Imaging*, vol. 23, pp. 661–675, 2004.
29. T. Damoulas and M. A. Girolami, "Probabilistic multi-class multi – Kernel learning: On protein fold recognition and remote homology detection," *Bioinformatics*, vol. 24, no. 10, pp. 1264–1270, 2008.
30. S. Amari and S. Wu, "Improving support vector machine classifiers by modifying kernel functions," *Neural Netw.*, vol. 6, pp. 783–789, 1999.
31. B. Souza and A. de Carvalho, "Gene selection based on multi-class support vector machines and genetic algorithms," *Mol. Res.*, vol. 4, no. 3, pp. 599–607, 2005.
32. R. O. Duda, P. E. Hart, and D. G. Stork, *Pattern classification*, 2nd ed. Hoboken, NJ: John Wiley & Sons Inc., 2001.
33. B. Schökopf and A. J. Smola, "Learning with kernels: Support vector machines, regularization, optimization, and beyond," *Adaptive computation and machine learning*. MIT press, 2002.
34. H. Fröhlich, O. Chapelle, and B. Scholkopf, "Feature selection for support vector machines by means of genetic algorithm," in *Proc. 15th. IEEE Intl. Conf. Tools with Artificial Intelligence*, pp. 142–148, 2003.
35. X. W. Chen, "Gene selection for cancer classification using bootstrapped genetic algorithms and support vector machines," in *Proc. IEEE Intl. Conf. Computational Systems, Bioinformatics Conference*, pp. 504–505, 2003.
36. C. Park and S. B. Cho, "Genetic search for optimal ensemble of feature-classifier pairs in DNA gene expression profiles," in *Proc. the Intl. Joint Conf. Neural Networks*, vol. 3, pp. 1702–1707, 2003.
37. F. A. Sadjadi, "Polarimetric radar target classification using support vector machines," *Opt. Eng.*, vol. 47, no. 4, pp. 046201, 2008.
38. H. Zhang, W. Huang, Z. Huang, and B. Zhang, "A kernel autoassociator approach to pattern classification," *IEEE Trans. Syst. Man Cyb. B*, vol. 35, no. 3, pp. 593–606, 2005.
39. W. S. Chen, P. C. Yuen, J. Huang, and D. Q. Dai, "Kernel machine-based one-parameter regularized Fisher discriminant method for face recognition," *IEEE Trans. Syst. Man Cyb. B*, vol. 35, pp. 659–669, 2005.
40. H. Xiong, M.N.S. Swamy, and M.O. Ahmad, "Optimizing the data-dependent kernel in the empirical feature space," *IEEE Trans. Neural Netw.*, vol. 16, pp. 460–474, 2005.

41. H. Xiong, Y. Zhang, and X. W. Chen, "Data-dependent kernel machines for microarray data classification," *IEEE/ACM Trans. Comput. Biol. Bioform.*, vol. 4, no. 4, pp. 583–595, 2007.
42. Y. Tan and J. Wang, "A support vector machine with a hybrid kernel and minimal Vapnik-Chervonenkis dimension," *IEEE Trans. Knowl. Data Eng.*, vol. 16, no. 4, pp. 385–395, 2004.
43. A. K. Jerebko, J. D. Malley, M. Franaszek, and R. M. Summers, "Multiple neural network classification scheme for detection of colonic polyps in CT colonography data sets," *Acad. Radiol.*, vol. 10, pp. 154–160, 2003.
44. J. Kivinen, A. J. Smola, and R. C. Williamson, "Online learning with kernels," *IEEE Trans. Signal Process.*, vol. 52, no. 8, pp. 2165–2176, 2004.
45. S. Ozawa, S. Pang, and N. Kasabov, "Incremental learning of chunk data for online pattern classification systems," *IEEE Trans. Neural Netw.*, vol. 19, no. 6, pp. 1061–1074, 2008.
46. Y. M. Li, "On incremental and robust subspace learning," *Pattern Recogn.*, vol. 37, no. 7, pp. 1509–1518, 2004.
47. Y. Kim, "Incremental principal component analysis for image processing," *Opt. Lett.*, vol. 32, no. 1, pp. 32–34, 2007.
48. T. Briggs and T. Oates, "Discovering domain specific composite kernels," in *Proc. the 20'th National Conf. Artificial Intelligence*, Pittsburgh, Pennsylvania; July 9–13, pp. 732–738, 2005.
49. N. Cristianini, J. Kandola, A. Elisseeff, and J. Shawe-Taylor, "On kernel target alignment," in *Proc. Neural Information Processing Systems*, Vancouver, British Columbia, Dec. 3–8, pp. 367–373, 2001.

5

CLOUD KERNEL ANALYSIS

5.1 INTRODUCTION

The most prominent limiting factor pertaining to widespread usage of the latest screening technology as a replacement for traditional screening colonoscopy is the limited supply of internal computed tomographic colonography (CTC) images available at a single medical institution. Such a limited supply of CTC training images significantly constrains the accuracy of an automated detection algorithm. As many medical institutions employ a limited number of CTC specialists, their computed tomography (CT) image analysis skills vary widely [1, 2]. To overcome these difficulties while providing a high-detection performance of polyps, computer-aided detection (CAD) schemes are investigated that semiautomatically detect suspicious lesions in CTC images [3].

CAD schemes have been developed for medical institutions (i.e., hospitals), where the number of training instances used during automated diagnosis is determined by resources and training requirements [4–8]. In most clinical settings, if more training data are collected after an initial tumor model is computed, retraining of the model becomes imperative in order to incorporate the newly added data from the local institution and to preserve the classification accuracy [4, 5, 9, 10]. This classification accuracy may be proportionately related to the amount of training data available, that is, more training data may yield a higher degree of classification [11, 12– 40]. With this in mind, we propose a new framework, called "Cloud Colonography," where the colonography database at each institution may be shared and/or uploaded to a single composite database on a cloud server. The CAD system at each institution can then be enhanced by incorporating new data from other institutions through the use of the distributed learning model proposed in this study.

Data-Variant Kernel Analysis, First Edition. Yuichi Motai.

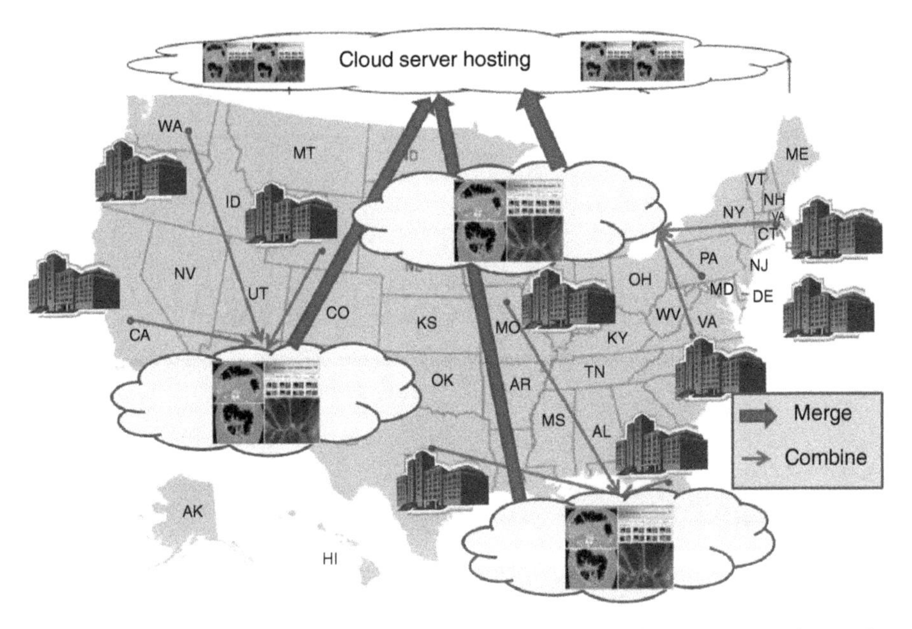

Figure 5.1 A concept of Cloud Colonography with distributed image databases for colon cancer diagnosis. The cloud server hosting will collect distributed databases from different institutions and group them using KA.

The concept of collaborative cloud applications using distributed databases has been discussed in [41–43], but not yet in the context of CAD in Cloud Colonography, meaning that the presented work is a first attempt at such a study. Therefore, the proposed Cloud Colonography framework shown in Fig. 5.1 requires a comprehensive study to determine whether the overall performance is improved by using multiple databases. However, it is worth noting that these existing studies showed that the overall classification performance for larger multiple databases was improved in practical settings [44].

The primary focus of this study is to develop Cloud Colonography using associated multiple databases (AMDs) to represent the statistical data characteristics in both private and public clouds. The proposed Cloud Colonography would be capable of learning multiple databases provided from multiple institutions. We propose composite kernel feature analysis with multiple databases for this specific problem. Kernel feature analysis is widely used for data compression, clustering/classification [45]. The improvement in performance could be obtained by employing such efficient cloud computing approaches in terms of resources, infrastructure, and operational costs. The contribution of this work is that the proposed Cloud Colonography applies AMD for practical clinical settings. The database acquired over a long period of time can sometimes be highly diverse, and each database is unique in nature; therefore, obtaining a clear distinction among multiple databases is a very challenging task [46, 47–63].

The rest of this chapter is organized as follows. Section 5.2 provides an introduction to cloud environment and a brief review of the existing feature extraction methods. Section 5.3 describes the proposed AMD for Cloud Colonography among multiple databases, Section 5.3.4 evaluates the experimental results for classification performance, Section 5.4.2 specifically accesses cloud computing performance, while conclusions are drawn in Section 5.5.4.

5.2 CLOUD ENVIRONMENTS

Cloud Colonography proposes new services using platform as a service (PaaS) models [42, 43]. Cloud Colonography is a cloud computing platform for programming, databases, and web servers. The clinical users access Cloud Colonography via the proposed software on a cloud platform without the cost and complexity of buying and managing the underlying hardware and software layers. Using existing PaaS platforms, such as Microsoft Azure and Google App Engine, the underlying computer and storage resources scale automatically to match application demand, so that the cloud user does not have to allocate resources manually.

Section 5.2.1 evaluates the existing cloud environment and effectiveness with respect to the proposed method. Section 5.2.2 introduces the relevant cloud framework, showing how the mathematical background on kernel principal component analysis (KPCA) is used for AMD.

5.2.1 *Server Specifications of Cloud Platforms*

The cloud vendor specification is shown in Table 5.1. In addition to private cloud environment using the intranet, the public cloud vendors using the Internet are considered to implement the proposed Cloud Colonography.

The leading commercial cloud hosting service, Amazon, is chosen because of its superiority of performance as listed in Table 5.1. Microsoft and IBM do not support MATLAB® programming, yet. Therefore, we decided to adopt Amazon as a public cloud platform for the implementation of Cloud Colonography. Private Cloud [64] is a department cluster used for a private cloud platform, and as a comparison, we added

Table 5.1 Representative cloud venders

Vendors	Monthly uptime (%)	Credit level max, %	RAM per 1 CPU, GB	APIs supports
Amazon [69]	99.95	30	0.615	REST, Java, IDE, Command line
Microsoft [74]	99.95	25	1.75	REST, Java, IDE, Command line
IBM [75]	99.00	10	2.00	REST
Private Cloud [64]	NA	NA	64	
Desktop PC	NA	NA	16	

a desktop PC as well. The effectiveness of Cloud Colonography mainly comes from the number of datasets that is manageable and the computational capacity of those large datasets in the programming. The proposed KPCA with AMD is a solution for these main problems. In the following subsection, we describe the mathematical background of KPCA in the cloud environment.

5.2.2 Cloud Framework of KPCA for AMD

For efficient feature analysis, extraction of the salient features of polyps is essential because of the huge data size from many institutions and the 3D nature of the polyp databases [3]. The desired method should allow for the compression of such big data. Moreover, the distribution of the image features of polyps is nonlinear [65]. To address these problems, we adopt the kernel approaches [45] in a cloud platform.

The key idea of kernel feature analysis is how to select a nonlinear, positive-definite kernel $K : R^d \times R^d \rightarrow R$ of an integral operator in the d-dimensional space. The kernel K, which is a Hermitian and positive semidefinite matrix, calculates the inner product between two finite sequences of inputs $\{x_i : i \in n\}$ and $\{x_j : j \in n\}$, defined as $K := (K(x_i, x_j)) = (\Phi(x_i).\Phi(x_j) : i, j \in n)$, where x is a gray-level CT image, n is the number of image database, and $\Phi : R^d \rightarrow H$ denotes a nonlinear embedding (induced by K) into a possibly infinite dimensional Hilbert space H as shown in Fig. 5.2. A more thorough discussion of kernels can be found in [66]. Our AMD module for CTC images is a dynamic extension of KPCA as follows.

KPCA uses a Mercer kernel [67] to perform a linear principal component analysis of the transformed image. Without loss of generality, we assume that the image of the data has been centered so that its scatter matrix in S is given by $S = \sum_{i=1}^{n} \Phi(x_i)(x_i)\Phi(x_i)^T$. The eigenvalues λ_j and eigenvectors e_j are obtained by solving the following equation, $\lambda_j e_j = Se_j = \sum_{i=1}^{n} \Phi(x_i)\Phi(x_i)^T e_j = \sum_{i=1}^{n} \langle e_j, \Phi(x_i)\rangle \Phi(x_i)$. If K is an $n \times n$ Gram matrix, with the element $k_{ij} = \langle \Phi(X_i), \Phi(X_j)\rangle$ and

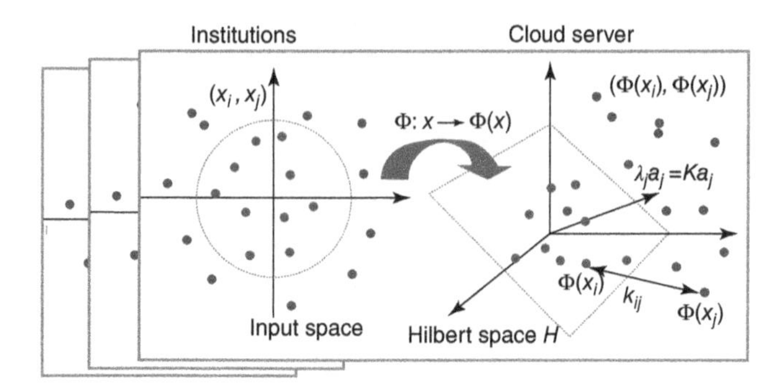

Figure 5.2 An illustration of KPCA mathematical background. KPCA calculates the eigenvectors and eigenvalues by analyzing the kernel feature space of multiple institutions so that a cloud server can handle larger datasets.

$a_j = [a_{j1}, a_{j2}, ..., a_{jn}]$ are the eigenvectors associated with eigenvalues λ_j, then the dual eigenvalue problem equivalent to the problem can be expressed as follows: $\lambda_j a_j = K a_j$.

A traditional KPCA can be extended into the cloud framework as follows:

1. Compute the Gram matrix that contains the inner products between pairs of image vectors.
2. Configure AMD by grouping datasets from multiple institutions.
3. Solve $\lambda_j a_j = K a_j$ to obtain the coefficient vectors a_j for $j = 1, \ 2, ..., n$.
4. The projection of a test point $x \in R^d$ along the jth eigenvector is $\langle e_j, \Phi(x) \rangle = \sum_{i=1}^{n} a_{ji} \langle \Phi(x_i), \Phi(x) \rangle = \sum_{i=1}^{n} a_{ji} k(x, \ x_i)$.

The above equation implicitly contains an eigenvalue problem of rank n, so the computational complexity of KPCA is $O(n^3)$. Each resulting eigenvector can be expressed as a linear combination of λ_* terms. The total computational complexity is given by $O(ln^2)$ where l stands for the number of features to be extracted and n stands for the rank of the Gram matrix K [66, 68]. Once we have our Gram matrix ready, we can apply these algorithms to our database to obtain a higher dimensional feature space. This cloud-based kernel framework will be discussed further in the following sections.

5.3 AMD FOR CLOUD COLONOGRAPHY

This section proposes the AMD method. Section 5.3.1 shows the overall concept of AMD for cloud environment, Section 5.3.2 explains more details on the design of AMD, Section 5.3.3 shows the two implementation cases for the selected public cloud environments, and Section 5.3.4 shows the proposed parallelization.

5.3.1 AMD Concept

As in Fig. 5.2, KPCA mentioned is executed for Cloud Colonography by analyzing the images of polyps with nonlinear big feature space. We apply KPCA to both an individual dataset and groups of datasets. KPCA can be used in conjunction with AMD to synthesize individual databases into larger composite databases shown in Fig. 5.3.

The concept of AMD is to format distinct distributed databases into a uniform larger database for the proposed Cloud Colonography platform. AMDs are expected to solve the issues of nonlinearity and excessive data size, so that CAD classification can achieve optimal performance. Specifically, kernels handle nonlinear to linear projection, and principal components analysis (PCA) handles data size reduction. The following subsection describes how our proposed kernel framework will be organized for AMD.

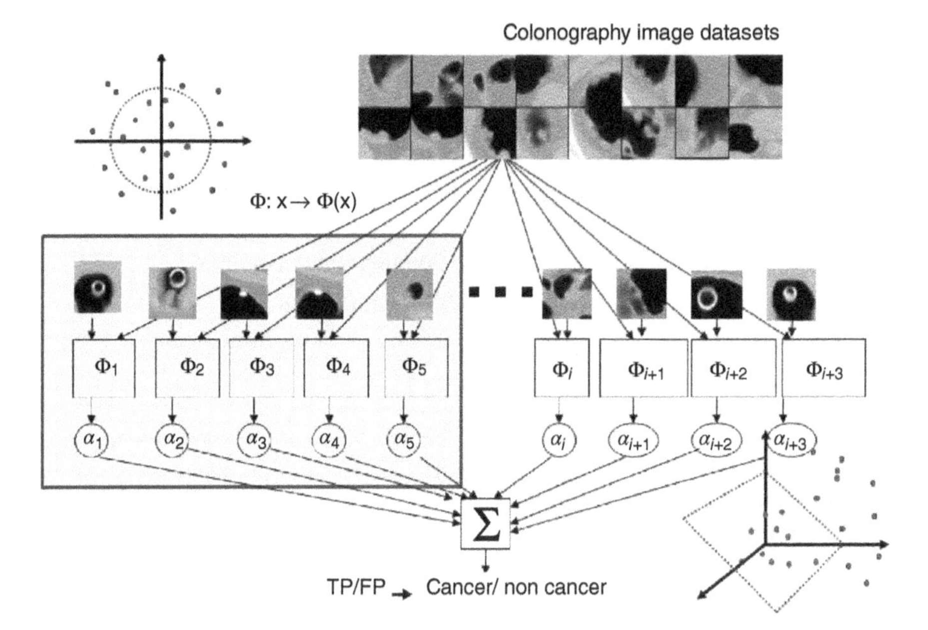

Figure 5.3 A concept of AMD. The proposed kernel framework combines the Cloud Colonography datasets by analyzing the images of polyps with nonlinear big feature space.

5.3.2 Data Configuration of AMD

We adapt Community Cloud [69] for Cloud Colonography to share infrastructure between several hospitals with common CTC domains [data compliance and security, Health Insurance Portability and Accountability Act (HIPPA) agreement, etc.]. The costs are spread over fewer users than a public cloud (but more than a private cloud), so only some cost savings are expected.

Figure 5.4 illustrates the AMD construction by analyzing the data from other individual institutional databases through the four representative steps. Step 1 is how to split each database, Step 2 is how to combine several databases, Step 3 is how to sort the combined databases, and Step 4 is how to merge the sorted databases. We explain individual steps as follows:

Step 1: Split We split each database by maximizing the Fisher scalar for kernel optimization [70]. The Fisher scalar is used to measure the class separability J of the training data in the mapped feature space. It is formulated as $J = \text{trace}\left(\sum_1 S_{bl}\right) / \text{trace}\left(\sum_1 S_{wl}\right)$, where S_{bl} represents "between-class scatter matrices" and S_{wl} represents "within-class scatter matrices." According to [71], the class separability by Fisher scalar can be expressed using the basic kernel matrix P^l (submatrices of ${P_{11}}^l$, ${P_{12}}^l$, ${P_{21}}^l$, and ${P_{22}}^l$) as $J_l(a_l) = a_l^T M_l a_l / a_l^T N_l a_l$, where $M_l = K_l^T B_l K_l$, $N_l = K_l^T W_l K_l$, and

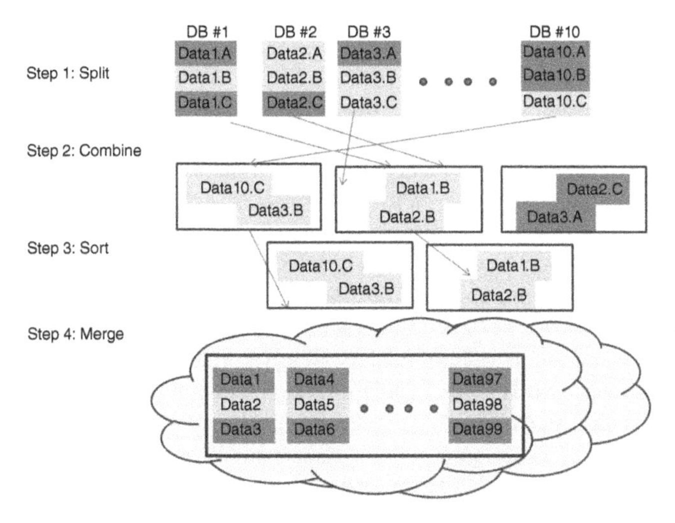

Figure 5.4 Four representative steps of AMD. The proposed AMD consists of the four main criteria to manage databases by (i) split, (ii) combine, (iii) sort, and (iv) merge.

$$W_l = \text{diag}(P^l_{11}, P^l_{22}) - \begin{pmatrix} P^l_{11}/n_1 & 0 \\ 0 & P^l_{22}/n_2 \end{pmatrix} \quad B_l = \begin{pmatrix} P^l_{11}/n_1 & 0 \\ 0 & P^l_{22}/n_2 \end{pmatrix} - \frac{P_l}{n} \tag{5.1}$$

To maximize $J_l(a_l)$, the standard gradient approach is followed. If the matrix N_{0i} is nonsingular, the optimal a_l that maximizes the $J_l(a_l)$ is the eigenvector that corresponds to the maximum eigenvalue of $M_l a_l = \lambda_l N_l a_l$. The criterion to select the best kernel function is to find the kernel that produces the largest eigenvalue.

$$\lambda_{l*} = \arg\max_{\lambda_l}(N_l^{-1} M_l) \tag{5.2}$$

Choosing the eigenvector that corresponds to the maximum eigenvalue can maximize the $J_l(a_l)$ to achieve the optimum solution. Once we determine the eigenvalues, the eigenvectors associated with the selected eigenvalues represent each split dataset.

Step 2: Combine Alignment is used to measure the adaptability of a kernel to the target data. Alignment provides a practical objective for kernel optimization whether datasets are similar or not. The alignment measure is defined as a normalized Frobenius inner product between the kernel matrix and the target label matrix introduced by Cristianini et al. [72]. The empirical alignment between the two kernels with respect to the training set S is given as

$$\text{Frob}(K_1, K_2) = \frac{\langle K_1, K_2 \rangle_F}{\|K_1\|_F \|K_2\|_F}$$

where K_i is the kernel matrix for the training set S, $\|K_i\|_F = \sqrt{\langle K_i, K_i \rangle_F}$, and $\langle K_i, K_j \rangle_F$ is the Frobenius inner product between K_i and K_j. It has been shown that if datasets chosen are well aligned with the other datasets, these datasets are combined.

Step 3: Sort We use class separability as a measure to identify whether the combined data is configured correctly in the right order. The separability within the combined datasets is required to check how well the data is sorted. It can be expressed as the ratio of separabilities: $\xi = J'_*(\alpha'_r)/J_*(\alpha_r)$ where $J'_*(\alpha'_r)$ denotes the class separability yielded by the most dominant kernel for the composite data (i.e., new incoming data and the previous offline data). $J_*(\alpha_r)$ is the class separability yielded by the most dominant kernel for another dataset. Thus, relative separability can be rewritten as: $\xi = \lambda'_*/\lambda_*$ where λ'_* corresponds to the most dominant eigenvalue of composite data to be tested, and λ_* is the most dominant eigenvalue of the combined/entire data. On the basis of the comparison of relative separabilities, the relationships among the combined datasets are finalized in the correct configuration. In this sorting step, we reduce the data size by ignoring the data with nondominant eigenvalues.

Step 4: Merge Finally, we consider merging the combined databases from Step 3.

Among the kernels k_i, $i = 1, 2,..., p$, we will tune ρ to maximize for the empirical alignment as follows:

$$\hat{\rho} = \arg_\rho \max(\text{Frob}(\rho, k_i, k_j)) = \arg\max \left(\frac{\left\langle \sum_i \rho_i K_i, K_j \right\rangle}{\sqrt{\left\langle \sum_i \rho_i K_i \right\rangle, \left\langle \sum_j \rho_i K_j \right\rangle}} \right)$$

$$= \arg\max_\rho \left(\frac{\rho^T V_{ij} \rho}{\rho^T U_{ij} \rho} \right) \tag{5.3}$$

where $u_i = \sqrt{\langle K_i, yy^T \rangle}$, $v_{ij} = \sqrt{\langle K_i, K_j \rangle}$, $U_{ij} = u_i u_j$, $V_{ij} = v_i v_j$, $\rho = (\sqrt{\rho_1}, \sqrt{\rho_2}, ..., \sqrt{\rho_p})$.

Let us introduce the generalized Raleigh coefficient, which is given as

$$J(\rho) = \frac{\rho^T V \rho}{\rho^T U \rho} \tag{5.4}$$

We obtain $\hat{\rho}$ by solving the generalized eigenvalue problem $V\rho = \delta U \rho$ where δ denotes the eigenvalues. Once we find this optimum composite coefficient $\hat{\rho}$, which will be the eigenvector corresponding to maximum eigenvalue δ, we can compute the composite data-dependent kernel matrix. Because we have associated all cloud data, which makes use of the data-dependent composite kernel starting from Steps 1 to 3, we can proceed to recombine all the institutions' data into small sets of merged databases, shown in Fig. 5.4.

AMD algorithm

Step 1 **Split** each database by maximizing the Fisher scalar.

Step 2 **Combine** other datasets if the alignment measure of datasets is high.

Step 3 **Sort** those combined datasets by class separability as a measure to identify whether the combined data is configured correctly in the right order.

Step 4 **Merge** the sorted cloud datasets from the institutions by computing the maximum eigenvalue δ for the composite data-dependent kernel matrix, which represents the associated datasets for KPCA.

The big advantage of the proposed AMD algorithm is to compress multiple datasets into the merged databases with the bounded data size. These compressed databases can be handled easier than the original databases, meaning a reduction in computational time, memory, and running cost.

5.3.3 Implementation of AMD for Two Cloud Cases

AMD's four steps are implemented into the two platform cases, private and public clouds, which are commercially available. We will demonstrate how the proposed Cloud Colonography is specifically adapted for this widespread implementation.

Case 1: Private Cloud The first specification of Cloud Colonography is listed in Fig. 5.5. This figure shows the representative layered implementation of the private cloud framework and its architectural components. A single institution/hospital handles the patient CTC datasets by a clinical IT staff or a third-party organization and hosts them either internally or externally. Self-run data centers/servers are generally capital intensive, requiring allocations of space, hardware, and environmental controls.

Case1 has a "layered architecture," with each layer adding further value and complimentary functionality to the data input from the layer below and providing the relevant output to the layer above. The solution architecture has the following four representative layers:

(*Layer1*) *AMD Monitoring/Management Layer.* This layer consists of various modules for CTC AMD, which monitors the characteristics of CTC databases as described in Section 5.3.2. These modules, as described in the AMD algorithm, generate a salient feature space, compare existing datasets to new ones, and prepare shared datasets for the next layer.

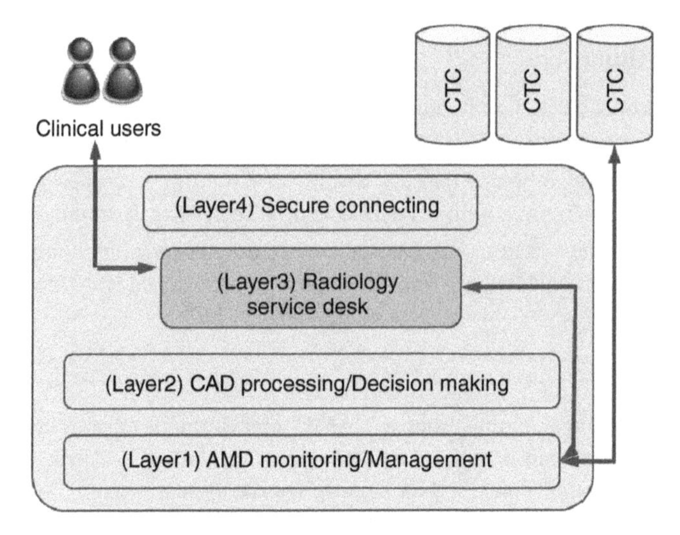

Figure 5.5 Cloud Colonography architecture in private cloud environment. The proposed Cloud Colonography consists of the four representative layers from CTC CAD analysis to the service desk reporting for clinical users.

(Layer2) CAD Processing/Decision-making Layer. All the CAD decision making from data collection of Layer1 is processed in this layer. This CAD processing includes the detection of cancer, such as patent and cancer identification for the diagnosis.

(Layer3) Radiology Service Desk. This layer further enables us to summarize the outcomes to increase radiological efficiency, such as visualization and annotation. These outcomes of the Service Desk provide views, reporting, administration, and operational support for practical and clinical uses.

(Layer4) Secure Connecting Layer. Qualified client–server applications are adapted for clinical users to assess Cloud Colonography. This layer is designed to prevent eavesdropping and tampering. On the ID request, this layer of the server switches the connection to Transport Layer Security (TLS) using a protocol-specific mechanism.

The private cloud specification is shown in Table 5.2 for the compassion of three representative platforms.

The computational performance is shown in Section 5.4.2, to analysis the speed, memory, and running cost. The private cloud server [64] has relatively large memory per CPU; however, the other specs of the personal desktop are superior to the private cloud. The personal desktop has the latest specification manufactured in 2014, and the private cloud server is 2 years older. The key specification is CPU clock speed; 3.6 GHz for the personal desktop, and 2.6 GHz for the private cloud. Other key

Table 5.2 Desktop and private cloud server hardware specification

Platform	Desktop computer	Private cloud server
Name	Dell Optiplex 9020	Dogwood
Processor name	Intel i7 4790	AMD Opteron 6282 SE
Clock Speed	3.6 GHz	2.6 GHz
# Processors	1	4
# Cores	4	16
L1 Cache	4 × 32KB Data 4 × 32KB Instruction	8 × 64 KB Data 16 × 16 KB Instruction
L2 Cache	4 × 256 KB	8 × 2 MB
L3 Cache, MB	8	2 × 8
Processor bus speed	5 GT/s DMI	3200 MHz 6.4 GT/s HTL
RAM	16 GB DDR3	64 GB per CPU
Memory bus speed, MHz	1600	1333
Storage technology	7.2k RPM SATA 3 Gbps	10k RPM SAS 6 Gbps

specifications of these servers for Cloud Colonography are the hard disk size of data storage and the random access memory (RAM) size.

Case 2: Public Cloud The cloud computing is also extended into a distributed set of servers that are running at different locations, while still connected to a single network or hub service as shown Fig. 5.6. Examples of this include distributed computing platforms such as BOINC [73], which is a voluntarily shared resource that implements cloud computing in the provisions model. A public cloud scenario is used for cloud computing when the services are rendered over a network that is open for multiple institutions/hospitals. Rather than a single database being used, the proposed AMD-based KPCA approach uses more than a single database to improve the cancer classification performance as shown in the experimental results.

Technically, there is little or no difference between the public and private cloud architecture; however, security consideration may be substantially different for services (applications, storage, and other resources) that are made available by a service provider for multiple institutions/hospitals. Generally, public cloud services provide a firewall, alike an extended Layer4 of private cloud scenario, to enhance a security barrier designed, and to prevent unauthorized or unwanted communications between computer networks or hosts. We adopt known cloud infrastructure service companies, such as Amazon AWS [69], Microsoft [74], and IBM [75] to operate the infrastructure at their data center for Cloud Colonography. The hardware specification of Amazon is shown in Table 5.3.

The computational performance is shown in Section 5.4.2, to analyze the speed, memory, and running cost for the Amazon EC2 instance servers, specifically c3.8xlarge for the experiments. Amazon has relatively large RAM and reasonable

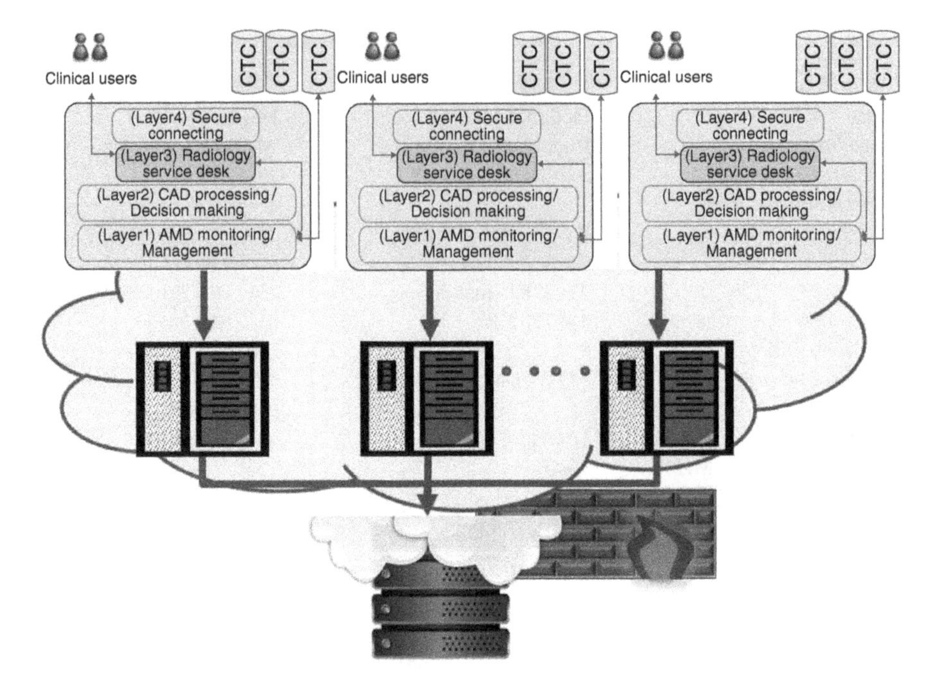

Figure 5.6 An extension of AMD framework in public cloud scenario.

pricing shown in Table 5.3. We have chosen Amazon because of MATLAB compatibility.

5.3.4 Parallelization of AMD

The parallelization programs for a public cloud server are proposed to optimize KPCA using MATLAB parallel toolbox. The parallel modules consist of (i) data parallelization of AMD and (ii) process parallelization of KPCA. MATLAB distributed computing servers use MATLAB job schedulers to distributed data and computational load to the cloud nodes. As Figure 5.7 shows, in [Module1], large-scaled CTC images are transformed into high-dimensional image-based feature space with the form of distributed array. In [Module2], those arrays are assigned and independently processed by multiple cores of each node. These two modules are fused by the head node in Fig. 5.7, which optimistically arranges the number of cores to the size of data array. To optimize the overall performance, we need two criteria: (i) minimizing inter-node data transmission for computational speed, and (ii) minimizing data storage for memory access requirement. The proposed criteria are designed to optimize the computational independency between nodes. The proposed method allows us to maximize the computational resources of elastic computing, which will reduce the computational time and required memory for the processing of pattern recognition algorithms.

Table 5.3 Amazon EC2 instance servers specification

Platform	c3.xlarge	c3.8xlarge	r3.xlarge	r3.8xlarge
Processor name	Intel Xeon E5-2670 v2	Intel Xeon E5-2670 v2	Intel Xeon E5-2670 v2	Intel Xeon E5-2670 v2
Clock Speed, GHz	2.5	2.5	2.5	2.5
# vCPU	4	4	4	32
# ECU	14	108	13	104
RAM, GB	7.5	60	30.5	244
Hard Disk Storage	80GB SSD	640GB SSD	80GB SSD	640GB SSD
Linux Instance Price, $/h	0.21	1.68	0.35	2.80
Processor Bus Speed	8GT/s DMI	8GT/s DMI	8GT/s DMI	8GT/s DMI
Memory Bus Speed, MHz	1866	1866	1866	1866

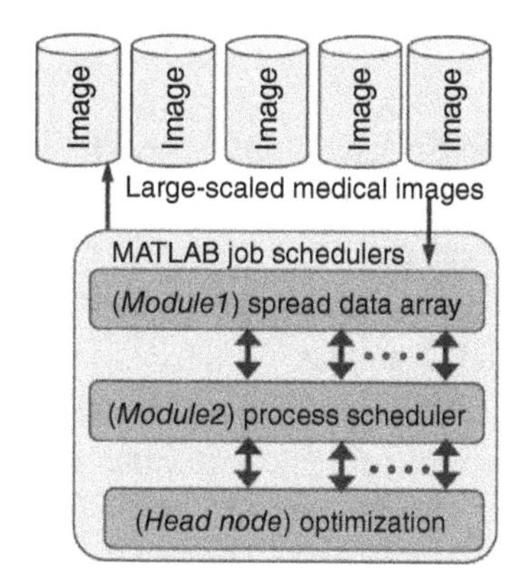

Figure 5.7 A proposed parallelization organization for cloud computing.

The proposed modules 1 and 2 are listed in the following pseudo codes. These steps are designed to reduce the computational time for the processing of pattern recognition algorithms.

Module 1: Pseudo algorithm

Data Parallelization

Input: new data, X_n

Parameter: Worker Node Structure, $\{DX,S\}$

Output: Update Distributed Dataset, DX

Begin

1. Vectorize multidimensional input: $V \leftarrow X_n$
2. Sort vector dataset according to dimension index
3. Slice V as S: $V_s \leftarrow \text{Reshape}(V,S)$
4. Append $V_s\{i\}$ at the end of $DX\{i\}$

End

Module 2: Pseudo algorithm

Training Process Parallelization

Input: Distributed Dataset

Parameter: Kernel parameter, Kernel Identifier

Output: Composite kernel Gram Matrix, Kernel Projector Model, K_c, Classifier Model

Begin

1. Locate and Load Node Data, $DX\{i\}$. Keep data private to the node.
2. Assign Node worker and Corresponding Data by data index.
3. Initiate cloud controller
4. Compute intermediate gram matrix according to kernel parameter
5. Terminate Cloud Controller
6. Return intermediate gram matrix to Head Node
7. Head Node use kernel parameter to combine intermediate gram matrix from individual workers and construct final gram matrix
8. Associate Gram Matrix with Class label, y
9. Find Eigen Components of the Gram Matrix and use Fisher Analysis to represent Eigen Vectors
10. Identify and Construct Kernel Projector Model, K_c
11. Initiate cloud Controller
12. Compute Kernel Projection using PCA
13. Terminate Cloud Controller
14. Compute Classifier Model

End

The proposed head node is designed to integrate the two parallelization algorithms. The following pseudo code shows testing the data samples using the modules 1 and 2.

Head node testing algorithm

Testing Process Parallelization
Input: Test Sample(s)
Parameter: Kernel Model, Kernel Projector Model
Output: Test decision Class
Begin

1. Use Data Parallelization Algorithm to distribute test samples to the workers
2. Compute Kernel Projection of the test samples
3. Employ Classifier model to test class of the test sample

End

5.4 CLASSIFICATION EVALUATION OF CLOUD COLONOGRAPHY

5.4.1 Databases with Classification Criteria

We used several cloud settings, using the database consisting of 464 CTC cases that were obtained from 19 institutions/hospitals in the United States and Europe. Our previously developed CAD scheme [3, 70, 76, 77] was applied to these CTC volumes, which yielded a total of 3774 detections (polyp candidates), consisting of 136 true positive (TP) detections and 3638 false positive (FP) detections. Note that the supine and prone CTC volumes of a patient were treated as independent in the detection process. A volume of interest (VOI) of 963 pixels was placed at each CTC candidate rather than the entire region of the CTC candidate. The collection of the VOIs for all of the candidates consisted of the databases used for the performance evaluation as shown in Table 5.4. We applied up to 40%-fold cross-variation for testing the training data. Note that the training and testing data were separated from the distinguished form. All the simulations were executed in a MATLAB environment optimized with the Parallel Computing Toolbox version [78]. The windows-operated desktop computing system featured an i7 processor with 3.6 GHz clock speed, and 16 GB of DDR3 memory.

Table 5.4 lists each database used for the classification results shown in Section 5.4.1. Using AMD, we connected these databases into assembled databases via cloud settings described in Section 5.3. The CAD classification results mainly come from the choice of databases, which means AMD chooses the databases used for assembling.

Table 5.5 shows the results of KPCA shown in Section 5.2 and AMD shown in Section 5.3. To assemble the databases, we applied KPCA to represent the databases. Using four steps of AMD, for example, Sigmoid and Gauss kernel functions represented the 12 databases (out of 19 databases) into three assembling databases (9-13-11-16, 19-6-121, 8-18-15-5) according to the order of the eigenvalue.

Table 5.4 Databases

	# Patients			# Database		
Databases	# TP patients	# FP patients	Total # patients	# TP	# FP	Total # database
1	5	30	35	12	155	167
2	3	29	32	5	217	222
3	3	10	13	7	213	220
4	3	27	30	5	206	211
5	7	35	42	12	196	208
6	3	25	28	6	198	204
7	3	24	27	6	208	214
8	1	28	29	4	200	204
9	3	17	20	8	190	198
10	3	22	25	7	198	205
11	4	23	27	8	181	189
12	3	29	32	4	191	195
13	2	11	13	8	208	216
14	3	27	30	5	188	193
15	3	15	18	7	147	154
16	3	5	8	8	221	229
17	3	12	15	7	169	176
18	2	25	27	12	169	181
19	2	11	13	5	183	188
Average	3.1	21.3	24.4	7.15	190.1	197.3

The second assembling databases were represented by Sigmoid and Poly for databases (2-3-14-17). The last assembling databases were Sigmoid and Linear group (4-7-10). We used Table 5.5 for the classification results in Section 5.4.1.

We demonstrated the advantage of Cloud Colonography in terms of the CAD classification performance in a more specific way by introducing numerical criteria. We evaluated the classification accuracy of CAD using receiver operating characteristic (ROC) of the sensitivity and specificity criteria as statistical measures of the performance. The true positive rate (TPR) defines a diagnostic test performance for classifying positive cases correctly among all positive instances available during the test. The false positive rate (FPR) determines how many incorrect positive results occur among all negative samples available during the test as follows:

$$\text{TPR} = \frac{\text{True positives (TP)}}{\text{True positives (TP)} + \text{False negatives (FN)}}$$

$$\text{FPR} = \frac{\text{False positives (FP)}}{\text{False positives (FP)} + \text{True negatives (TN)}}$$

We also evaluated the alternative numerical values, area under the curves (AUC) by $\text{AUC} = \sum_{i=2}^{Z}(\text{FPR}_i - \text{FPR}_{i-1})\text{TPR}_i$, where Z is the number of discrete FPR_i. The proposed statistical analysis by use of AMD was applied to the multiple databases in Table 5.4, which showed that the CTC data were highly biased toward FPs

Table 5.5 AMD with KPCA for assembled database

Kernel type	Database assembling	First kernel	Second kernel
Sigmoid and Gauss	Database9	Sigmoid	Gauss
	Database13	$d = 6.67 * 10^{-4}$ Offset = 0	$\sigma = 0.4$
	Database11	$\rho_1 = 0.49$	$\rho_2 = 0.51$
	Database16		
	Database19	Sigmoid	Gauss
	Database6	$d = 8.89 * 10^{-4}$ Offset = 0	$\sigma = 0.2$
	Database12	$\rho_1 = 1.00$	$\rho_2 = 0.12$
	Database1		
	Database8	Sigmoid	Gauss
	Database18	$d = 7.78 * 10^{-4}$ Offset = 0	$\sigma = 0.3$
	Database15	$\rho_1 = 0.95$	$\rho_2 = 0.05$
	Database5		
Sigmoid and poly	Database2	Sigmoid	Linear
	Database3	$d = 3.34 * 10$ Offset = 0	$d = 3$, Offset = 0.1
	Database14	$\rho_1 = 0.71$	$\rho_2 = 0.20$
	Database17		
Sigmoid and linear	Database4	Sigmoid	Linear
	Database7	$d = 3.34 * 10$ Offset = 0.	$d = 13$
	Database10	$\rho_1 = 1.00$	$\rho_2 = 0.12$

(the average ratio between TP and FP is 1 : 26.6) due to the limited number of TPs caused by asymptomatic patient cohort.

5.4.2 Classification Results

We used the k-nearest neighbor (K-NN) method for Table 5.4 for classification with the parameter k, accordingly to show the performance of TPR and specificity with respect to the variable k in Figs. 5.8–5.10.

Figure 5.8 shows the ROC results for the Sigmoid and Gauss group kernels. We compared assembling databases performance, with a single database (databases 9, 11, 13, and 16) in Fig. 5.8(a), a single database (databases 1, 6, 12, and 19) in Fig. 5.8(b), and a single database (databases 5, 8, 15, and 18) in Fig. 5.8(c), respectively. In Fig. 5.8(a), AMD for assembling database outperformed the KPCA for a single database by achieving a higher TPR and a lower FPR. In Fig. 5.8(b), although Database 1 had a smaller FPR, AMD for assembling database had better gradient. In Fig. 5.8(c), the performance of AMD for assembling database was not as good as KPCA for database 18, but it performed better than the other three databases in terms of the ROC.

Figure 5.9 shows that KPCA with database 17 had the best performance in ROC, and the second best was the performance of AMD for assembling database. Regarding KPCA for databases 2, 3, and 14, FPR was a bit high, and TPR was not good enough at the same time. In the Sigmoid and poly group kernel experiment, AMD for assembling databases was not the best one, but better than KPCA for the rest of the single databases.

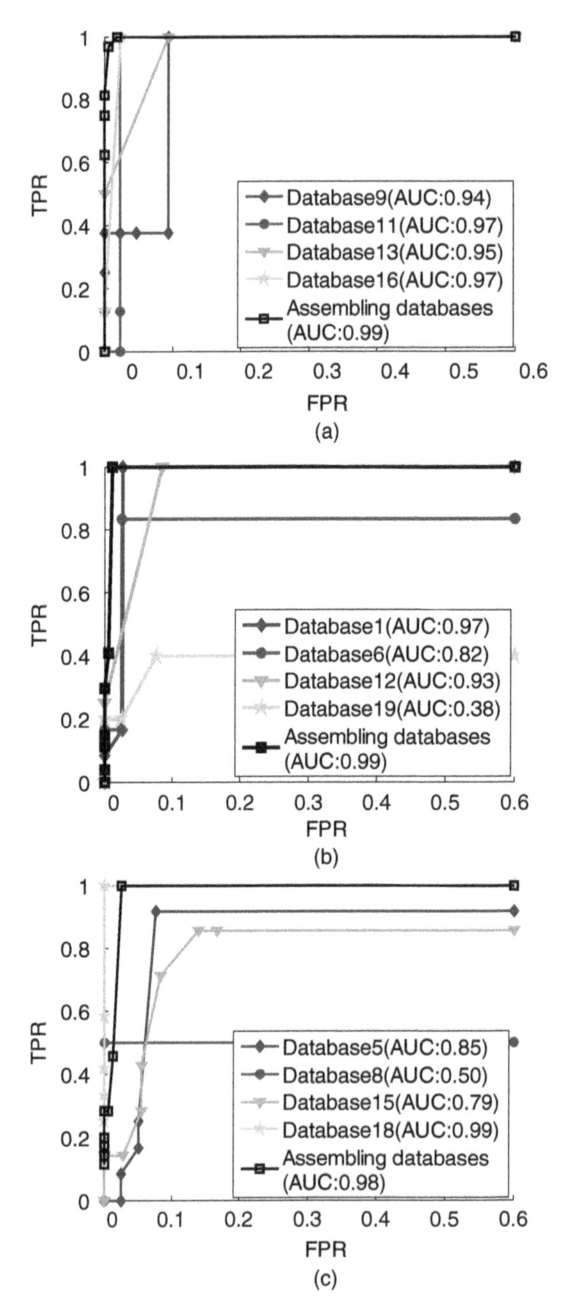

Figure 5.8 ROC with Sigmoid and Gauss group kernel. (a) Comparison between assembling databases and databases 9, 11, 13, and 16. (b) Comparison between assembling databases and databases 1, 6, 12, and 19. (c) Comparison between assembling databases and databases 5, 8, 15, and 18.

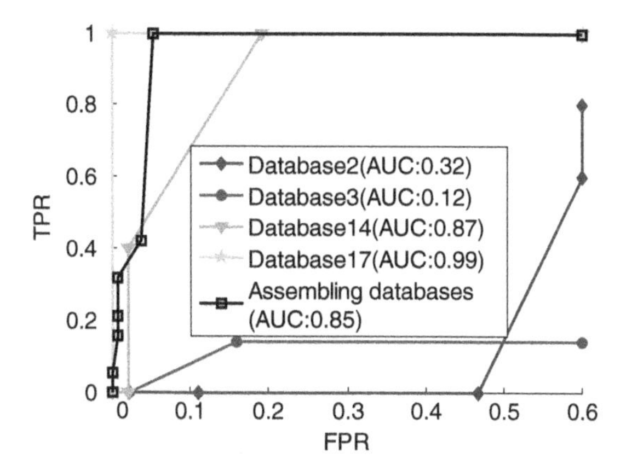

Figure 5.9 ROC with Sigmoid and poly group kernel.

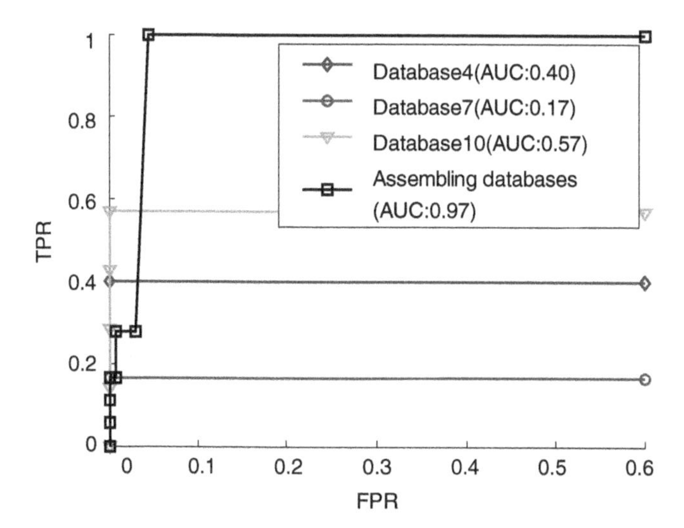

Figure 5.10 ROC with Sigmoid and linear group kernel.

In Fig. 5.10, we can see that AMD for assembling database outperformed KPCA for databases 4, 7, and 10, although database 10 had small FPR, but the maximal TPR only reached 0.58. On the other hand, AMD for assembling databases reached 0.98, although FPR was bigger than Database 10. In most cases, AMD for assembling databases has the advantage over KPCA by an average of 22.2% in TPR and 3.8% in FPR for a single database with respect to FPR and TPR from Figs. 5.8–5.10.

These ROC results show the calculated AUC (vertical axis) for the proposed method as the total network data size (horizontal axis) increases (mostly FT data). These figures show that Cloud Colonography with AMD exhibited AUC performance for classification in handling the AMD in all five different cloud environments

Table 5.6 Classification accuracy for assembled database

Kernel type	Database assembling	Dataset size, MB	Accuracy, %
Sigmoid and Gauss	Database9	916	97.54
	Database13	988	
	Database11	868	
	Database16	1064	
	Database19	780	100.00
	Database6	940	
	Database12	900	
	Database1	780	
	Database8	932	97.94
	Database18	756	
	Database15	696	
	Database5	960	
Sigmoid and poly	Database2	1028	95.33
	Database3	1036	
	Database14	876	
	Database17	820	
Sigmoid and linear	Database4	976	97.35
	Database7	251	
	Database10	956	

shown in Section 5. The ability to track changes using growing database size was also verified by the results shown in the following Section.

Table 5.6 shows that the proposed Cloud Colonography with AMD, consistently performed well with high classification accuracy, ranging from 95.33% to 100%. In general, there are no significant differences of cloud settings for classification accuracy. As the data size used increased, other performance criteria such as computational load and resources affect the overall system performance. In the following section, the cloud environments were analyzed for their efficiency in Cloud Colonography.

5.5 CLOUD COMPUTING PERFORMANCE

We evaluated private and public cloud scenarios in four aspects of the proposed design in Cloud Colonography. These are examined in Section 5.5, how databases are formed into AMD, and Sections 5.5.1–5.5.5.

5.5.1 Cloud Computing Setting with Cancer Databases

Table 5.7 shows how databases are formed into AMD. Three databases were generated from Table 5.4 for the analysis of both private and public cloud environments. These three datasets are a synthesis of the existing 19 databases by merging them into the cloud server; thus, the classification performance was equivalent to

Table 5.7 Databases for cloud computing analysis

	Training datasets				Testing datasets			
Databases	#TP	#FP	#Total	Size, MB	#TP	#FP	#Total	Size, MB
1	122	1,289	1,411	9,524	14	157	171	1,154
2	120	853	973	6,568	16	110	126	851
3	121	1,071	1,192	8,046	15	133	148	999
Total	363	3,213	3,576	24,138	45	400	445	3,004

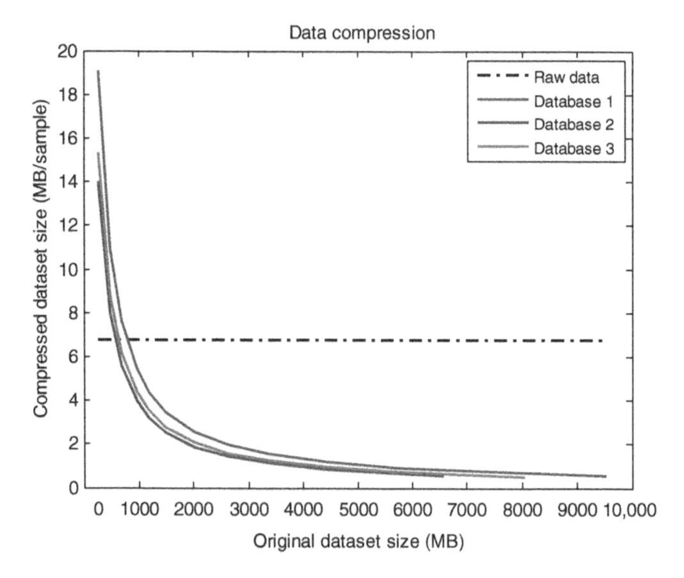

Figure 5.11 Data compression ratio for data size. The horizontal axis denotes the size of the data, and vertical axis denotes the compressed data ratio. In the three databases, the data compressions are relatively distinct to each other.

Section 5.3.4. These three generated databases are mainly used for the evaluation of cloud computing.

Figure 5.11 shows the compressed data ratio (vertical axis) of the Cloud Colonography network as the total network data size (horizontal axis) increases. The three databases for cloud environments are used to analyze how the proposed method handles the enlarged data sizes. The data compression ratio was defined as the size of the AMD feature space versus that of the raw data. As the data size increased, the compressed data ratio was reduced. Compared to the raw data, the three generated databases worked well to maintain the size for the analysis over the entire data size using AMD. The compression ratio reflects the corresponding decrease in heterogeneous to homogenous data. In other words, a constant ratio with increasing data size indicates an equal growth in both heterogeneous and homogenous data, such as in the case of raw data. The evaluation criteria for the experimental datasets start from computational speed with varying database sizes. These results have been fully outlined in the remaining sections that follow.

5.5.2 Computation Time

In this subsection, we analyzed the time required for processing of Cloud Colonography with AMD. All the experiments were implemented in MATLAB R2010a using the Statistical Pattern Recognition Toolbox for the Gram matrix calculation and kernel projection. For processing the large volume of data, private cloud environments were used. Run time was determined using the cputime command, representing how significant computational savings were for each environment.

Figure 5.12 shows the total computational time required for growing network database sizes in Amazon EC2 (as a public cloud), the Private Cloud and Desktop environment. The total training time was measured for three uncompressed databases listed in Table 5.7. The total training time was increased as the number of samples increased. The difference between the three environments was relatively small.

Figure 5.13 shows the mean execution time for compressed datasets. Compared to Fig. 5.12, the computation time was much improved when the datasets were compressed. The Public and Private Cloud required more computation time than the Desktop for all three data cases. The time difference increased as the number of datasets increased. The difference was calculated by averaging three databases shown in Table 5.8.

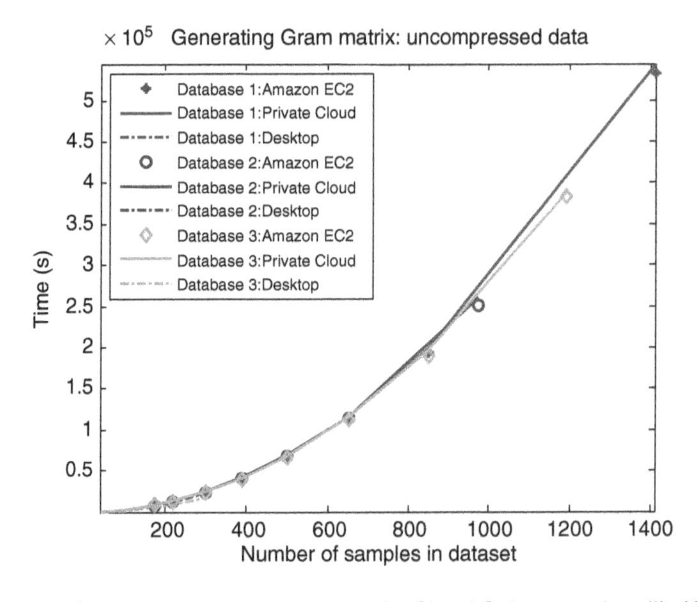

Figure 5.12 Total training time required for Cloud Colonography with AMD.

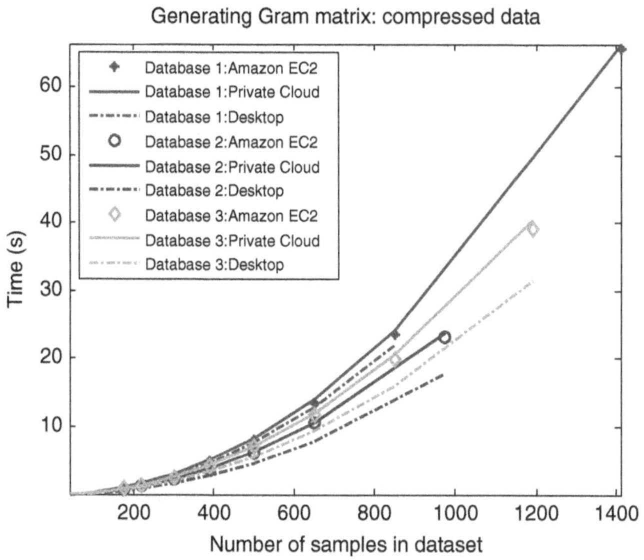

Figure 5.13 Mean execution time for Cloud Colonography with AMD.

Table 5.8 Averaged training computation time for three databases

Cloud	Uncompressed data, s	Compressed data, s
Amazon EC2	$1.4*10^5$	14.6
Private cloud	$1.4*10^5$	14.6
Desktop	$1.4*10^5$	12.3

Table 5.8 shows the average of the total training time and mean execution time shown in Figs. 5.12 and 5.13. These values were calculated by averaging three databases for each cloud environment. Table 5.8 demonstrates that the computation time for the public and private cloud was 18% larger than the desktop, meaning that the desktop was 18%, on average, faster than the private cloud. On the basis of the hardware specification in Table 5.2, the CPU speed of the desktop was 38% faster than the private cloud, and 44% faster than the public cloud (Amazon EC2 r3.8xlarge). The difference of the computational time between uncompressed and compressed datasets was over 10^4. The big reduction of computational time was achieved by the AMD due to the data compression. The increased ratio of the computation time in Figs. 5.12 and 5.13 shows that the proposed method was computationally efficient as the overall network database size increased. Therefore, Cloud Colonography was better suited to handle long-term sequences in a practical manner. Our computational speed for larger CTC databases is promising even if much larger datasets are used for the screenings.

Figure 5.14 selected the main module for the training time to specifically calculate Gram matrix in Cloud Colonography with AMD. The calculation of Gram matrix is

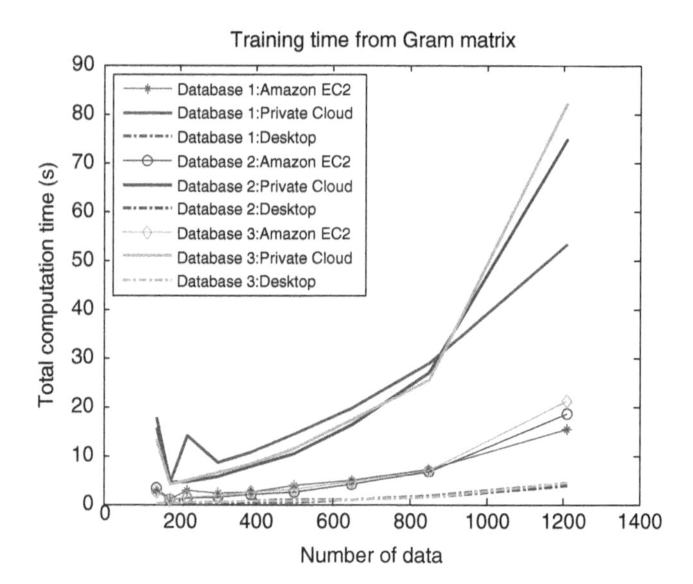

Figure 5.14 Computational time for Gram matrix in Cloud Colonography with AMD.

Table 5.9 Average total testing time per sample

Database	Amazon EC2, ms	Private cloud, ms	Desktop environment, ms
1	361	2851.4	361
2	263	2037.1	263
3	289.7	2232.9	289.7

the key module to computer KPCA with comparison between the Desktop and private cloud. Figure 5.14 also shows that Desktop was faster than public and private cloud for all three databases for the module of Gram matrix. The public and private cloud required more computation time than the Desktop for testing phase as well as shown in Table 5.9.

5.5.3 Memory Usage

In this subsection, we examine the degree of memory usage in Cloud Colonography with AMD as the data size changes. Figure 5.15 illustrates the effect of continuously increasing data size on the observed level of memory usage in each of the three databases.

Figure 5.15 shows that the proposed framework required more memory usage as the training sample increased in size. There are no differences of memory usage among cloud environments examined in this study. The degree of memory usage was proportionally increased as the training samples increased. We needed to develop more efficient method of memory usage, as the database size grew.

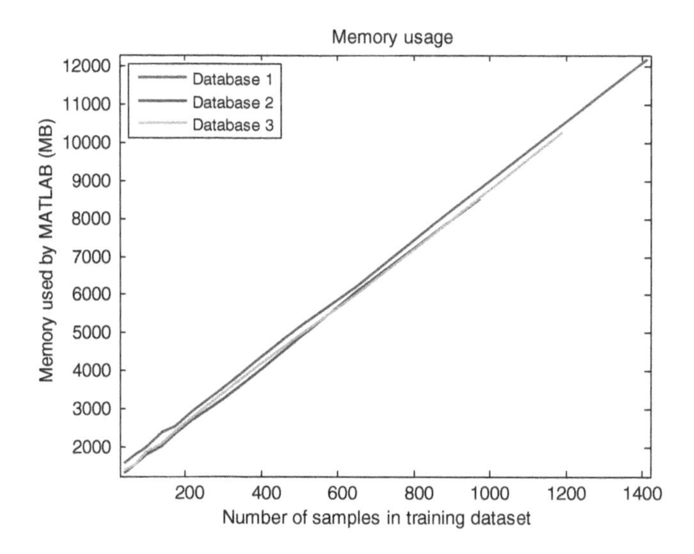

Figure 5.15 Memory usage for Cloud Colonography with AMD.

5.5.4 Running Cost

Cloud service providers typically charge on the basis of usage of memory, computation amount, storage, data transfer speed, and subscription type of the service. The Private Cloud service, as well as Desktop, was offered free of charge for this research purpose, but this was not the case for the public cloud. Thus, we estimated the cost on the basis of resource consumption for the three databases used.

Tables 5.10 and 5.11 illustrate the total cost component of the total storage data, memory, and processing in the cloud for three different databases used in this study. The result in previous sections, such as Fig. 5.15, showed that the total cost for cloud data storage increased in proportion to the total data stored on the cloud. As the size of the data stored was increased, the training time was also increased, which added to the total cost.

5.5.5 Parallelization

The proposed parallelization method described in Section 5.3.4 was tested on both private and public cloud environment. Computation time for large uncompressed

Table 5.10 Cost components for private cloud service

Cost component	Database 1	Database 2	Database 3
Storage, GB	12.3	8.9	10.7
Training memory, MB	12.2×10^3	10.28×10^3	8.523×10^3
Testing memory, MB	2.12×10^3	1.61×10^3	1.81×10^3
Training processing, trillion FLOPs	907	436	648
Testing processing, trillion FLOPs	12.8	6.9	8.9

Table 5.11 Cost components for public cloud Amazon EC2 (c3.8xlarge)

Cost component	Database 1	Database 2	Database 3
Storage, GB	12.3	8.9	10.7
Maximum training memory, MB	14.5×10^3	11.0×10^3	12.9×10^3
Maximum testing memory, MB	2.16×10^3	1.64×10^3	1.85×10^3
Approximate marginal cost, $	9.25	4.40	6.60

Table 5.12 Time and memory using parallelization for private cloud

Module environment	Worker	Computation time mean (ms/sample)	Maximum memory, GB
Parallelization	16	0.519	~59
	32	0.271	~63
Without parallelization	1	78.24	~35

Table 5.13 Time and cost using parallelization for public cloud Amazon EC2 (c3.8xlarge)

Module environment	Node	Worker	Computation time mean, ms/sample	Approximate cost, $
With parallelization	1	16	0.504	2.75
	2	32	0.268	2.05
	4	64	0.145	1.91
Without parallelization	1	1	76.82	6.60

unified Database (~25 GiB) experienced approximately 100-fold improvement. The summary of computation time for training from uncompressed data is summarized in Table 5.12 for private cloud and in Table 5.13 for public cloud.

As shown in Tables 5.12 and 5.13, the computational time was dramatically improved if the proposed parallelization module was implemented. In the case of public cloud, the cost was reduced by more than half. These results demonstrate that our proposed parallelization is effective for Cloud Colonography.

5.6 CONCLUSIONS

We proposed a new framework of Cloud Colonography, using different types of cloud computing environments. The databases from the CTC screening tests among several hospitals are going to be networked in the near future via cloud computing technologies. The proposed method called AMD was developed in this study for handling multiple CTC databases. When AMD is used for assembling databases, it can achieve almost 100% classification accuracy. The proposed AMD has the potential to play as a core classifier tool in the cloud computing framework for a model-based CAD scheme, which will yield high detection performance of polyps using KPCA for multiple institutions databases. Two cases in the proposed cloud platform are private and public. The computation time, memory usage, and running costs were compared between private and public cloud environments. The parallelization was successful developed to reduce the speed and cost. CTC CAD in the cloud computing environment may advance the clinical implementation of cancer screening and promote the early diagnosis of colon cancer.

REFERENCES

1. S. Winawer, R. Fletcher, D. Rex, J. Bond, R. Burt, J. Ferrucci, T. Ganiats, T. Levin, S. Woolf, D. Johnson, L. Kirk, S. Litin, and C. Simmang, "Colorectal cancer screening and surveillance: Clinical guidelines and rationale-Update based on new evidence," *Gastroenterology*, vol. 124, pp. 544–560, 2003;b*Systems*, vol. 36, no. 1, pp. 266–278, 2000.
2. J. G. Fletcher, F. Booya, C. D. Johnson, and D. Ahlquist, "CT colonography: Unraveling the twists and turns," *Curr. Opin. Gastroenterol.*, vol. 21, pp. 90–98, 2005.
3. H. Yoshida and A. H. Dachman, "CAD techniques, challenges and controversies in computed tomographic colonography," *Abdom Imaging*, vol. 30, pp. 26–41, 2005.
4. G. Kiss, J. Van Cleynenbreugel, M. Thomeer, P. Suetens, and G. Marchal, "Computer-aided diagnosis in virtual colonography via combination of surface normal and sphere fitting methods," *Eur. Radiol.*, vol. 12, pp. 77–81, 2002.
5. D. S. Paik, C. F. Beaulieu, G. D. Rubin, B. Acar, R. B. Jeffrey, Jr., J. Yee, J. Dey, and S. Napel, "Surface normal overlap: A computer-aided detection algorithm with application to colonic polyps and lung nodules in helical CT," *IEEE Trans. Med. Imaging*, vol. 23, pp. 661–675, 2004.
6. J. Näppi, H. Frimmel, A. H. Dachman, and H. Yoshida, "A new high-performance CAD scheme for the detection of polyps in CT colonography," in *Med Imaging 2004: Image Processing*, San Diego, CA, February 14, pp. 839–848, 2004.
7. A. K. Jerebko, J. D. Malley, M. Franaszek, and R. M. Summers, "Multiple neural network classification scheme for detection of colonic polyps in CT colonography databases," *Acad. Radiol.*, vol. 10, pp. 154–160, 2003.
8. A. K. Jerebko, J. D. Malley, M. Franaszek, and R. M. Summers, "Support vector machines committee classification method for computer-aided polyp detection in CT colonography," *Acad. Radiol.*, vol.12, pp. 479–486, 2005.
9. R. M. Summers, C. F. Beaulieu, L. M. Pusanik, J. D. Malley, R. B. Jeffrey, Jr., D. I. Glazer, and S. Napel, "Automated polyp detector for CT colonography: Feasibility study," *Radiology*, vol. 216, pp. 284–290, 2000.
10. R. M. Summers, M. Franaszek, M. T. Miller, P. J. Pickhardt, J. R. Choi, and W. R. Schindler, "Computer-aided detection of polyps on oral contrast-enhanced CT colonography," *AJR Am. J. Roentgenol.*, vol. 184, pp. 105–108, 2005.
11. Z. B. Xu, M. W. Dai, and D. Y. Meng, "Fast and efficient strategies for model selection of gaussian support vector machine," *IEEE Trans. Syst. Man Cyb. B*, vol. 39, pp. 1292–1207, 2009.
12. M. Doumpos, C. Zopounidis, and V. Golfinopoulou, "Additive support vector machines for pattern classification," *IEEE Trans. Syst. Man Cyb. B* , vol.37, pp. 540–550, 2007.

13. C. F. Juang, S. H. Chiu, and S. J. Shiu, "Fuzzy system learned through fuzzy clustering and support vector machine for human skin color segmentation," *IEEE Trans. Syst. Man Cyb. A*, vol.37, pp. 1077–1087, 2007.
14. Y. Shao, and C. H. Chang, "Two dimensional principal component analysis based independent component analysis for face recognition," *IEEE Trans. Syst. Man Cyb. BA*, vol. 41, pp. 284–293, 2011.
15. M. Xu, P. M. Thompson, and A. W. Toga, "Adaptive reproducing kernel particle method for extraction of the cortical surface," *IEEE Trans. Med. Imaging*, vol. 25, no. 6, pp. 755–762, 2006.
16. I. E. A. Rueda, F. A. Arciniegas, and M. J. Embrechts, "Bayesian separation with sparsity promotion in perceptual wavelet domain for speech enhancement and hybrid speech recognition," *IEEE Trans. Syst. Man Cyb. A*, vol.34, pp. 387–398, 2004.
17. W. Zheng, C. Zou, and L. Zhao, "An improved algorithm for kernel principal component analysis," *Neural Process Lett.*, pp. 49–56, 2005.
18. J. Kivinen, A. J. Smola, and R. C. Williamson, "Online learning with kernels," *IEEE Trans. Signal Process.*, vol. 52, no. 8, pp. 2165–2176, 2004.
19. S. Ozawa, S. Pang, and N. Kasabov, "Incremental learning of chunk data for online pattern classification systems," *IEEE Trans. Neural Netw.*, vol. 19, no.6, pp. 1061–1074, 2008.
20. H. T. Zhao, P. C. Yuen, and J. T. Kwok, "A novel incremental principal component analysis and its application for face recognition," *IEEE Trans. Syst. Man Cyb. B* , vol. 36, no. 4, pp. 873–886, 2006.
21. Y. M. Li, "On incremental and robust subspace learning," *Pattern Recogn.*, vol. 37, no. 7, pp. 1509–1518, 2004.
22. Y. Kim, "Incremental principal component analysis for image processing," *Opt. Lett.*, vol. 32, no. 1, pp. 32–34, 2007.
23. B. J. Kim, and I. K. Kim, "Incremental nonlinear PCA for classification," in *Proc. Knowledge Discovery in Databases*, Pisa, Italy, September 20–24, vol. 3202, pp. 291–300, 2004.
24. B. J. Kim, J. Y. Shim, C. H. Hwang, I. K. Kim, and J. H. Song. "Incremental feature extraction based on empirical kernel map," *Found. Intell. Syst.*, vol. 2871, pp. 440–444, 2003.
25. L. Hoegaerts, L. De Lathauwer, I. Goethals, J. A. K. Suykens, J. Vandewalle, and B. De Moor. "Efficiently updating and tracking the dominant kernel principal components," *Neural Netw.*, vol. 20, no. 2, pp. 220–229, 2007.
26. T. J. Chin, and D. Suter. "Incremental kernel principal component analysis," *IEEE Trans. Image Process.*, vol. 16, no. 6, pp. 1662–1674, 2007.
27. V. N. Vapnik, *The nature of statistical learning theory*, 2nd ed. New York: Springer, 2000.
28. T. Damoulas and M. A. Girolami, "Probabilistic multi-class multi-kernel learning: On protein fold recognition and remote homology detection," *Bioinformatics*, vol. 24, no. 10, pp. 1264–1270, 2008.
29. S. Amari and S. Wu, "Improving support vector machine classifiers by modifying kernel functions," *Neural Netw.*, vol.6, pp. 783–789, 1999.
30. B. Souza and A. de Carvalho, "Gene selection based on multi-class support vector machines and genetic algorithms," *Mol. Res.*, vol. 4, no.3, pp. 599–607, 2005.
31. H. Xiong, Y Zhang and X. W. Chen, "Data-dependent kernel machines for microarray data classification," *IEEE/ACM Trans. Comput. Biol. Bioinform.*, vol.4, no. 4, pp. 583–595, 2007.

32. H Fröhlich, O Chapelle and B Scholkopf, "Feature selection for support vector machines by means of genetic algorithm," in *Proc. 15th IEEE Int. Conf. Tools with Artificial Intelligence*, Dayton, OH, November 3–5, pp. 142–148, 2008.
33. X. W. Chen, "Gene selection for cancer classification using bootstrapped genetic algorithms and support vector machines," in *Proc. IEEE Int. Conf. Computational Systems, Bioinformatics Conference*, Stanford, California, August 11–14, pp. 504–505, 2003.
34. C. Park and S. B. Cho, "Genetic search for optimal ensemble of feature-classifier pairs in DNA gene expression profiles," in *Proc. Int. Joint Conf. Neural Networks*, Portland, Oregon, July 20–24, vol. 3, pp. 1702–1707, 2003.
35. F. A. Sadjadi, "Polarimetric radar target classification using support vector machines," *Opt. Eng.*, vol. 47, no. 4, 2008.
36. C. Hu, Y. Chang, R. Feris, and M. Turk, "Manifold based analysis of facial expression," in *Computer Vision and Pattern Recognition Workshop*, Washington, DC, USA, June 27 2004–July 2, vol. 27, pp. 81–85, 2004.
37. T. T. Chang, J. Feng, H. W. Liu, and H. Ip, "Clustered microcalcification detection based on a multiple kernel support vector machine with grouped features," in *Proc 19th Int. Conf. Pattern Recognition*, Tampa, FL, December 8–11, vol. 8, pp. 1–4, 2008.
38. T. Briggs, and T. Oates, "Discovering domain specific composite kernels," *Proc 20th national conf. artificial Intelligence*, Pittsburgh, Pennsylvania, July 9–13, pp. 732–738, 2005.
39. G. R. G. Lanckriet, N. Cristianini, P. Bartlett, L. Ghaoui, and M. I. Jordan, "Learning the kernel matrix with semidefinite programming," *J. Mach. Learn. Res.*, vol. 5, pp. 27–72, 2004.
40. Y. Tan, and J. Wang, "A support vector machine with a hybrid kernel and minimal Vapnik-Chervonenkis dimension," *IEEE Trans. Knowl. Data Eng.*, vol. 16, no. 4, pp. 385–395, 2004.
41. S. Di, C. Wang and F. Cappello, "Adaptive algorithm for minimizing cloud task length with prediction errors," *IEEE Trans. Cloud Comput.*, vol. 2, no. 2, pp. 194–207, 2014.
42. C. Tsai, W. Huang, M. Chiang, M. Chiang and C. Yang, "A hyper-heuristic scheduling algorithm for cloud," *IEEE Trans. Cloud Comput.*, vol. 2, no. 2, pp. 236–250, 2014.
43. F. Larumbe and B. Sanso, "A tabu search algorithm for the location of data centers and software components in green cloud computing networks," *IEEE Trans. Cloud Comput.*, vol. 1, no. 1, pp. 22–35, 2013.
44. J. H. Yao, M. Miller, M. Franaszek, and R.M. Summers, "Colonic polyp segmentation in CT colonography-based on fuzzy clustering and deformable models," *IEEE Trans. Med. Imaging*, vol. 23, pp. 1344–1356, 2004.
45. Y. Motai and H. Yoshida, "Principal Composite Kernel Feature Analysis: Data-Dependent Kernel Approach," *IEEE Trans. Knowl. Data Eng.*, vol.25, no.8, pp. 1863–1875, 2013.
46. A. K. Jerebko, R. M. Summers, J. D. Malley, M. Franaszek, and C. D. Johnson, "Computer-assisted detection of colonic polyps with CT colonography using neural networks and binary classification trees," *Med. Phys.*, vol. 30, pp. 52–60, 2003.
47. A. K. Jerebko, R. M. Summers, J. D. Malley, M. Franaszek, and C. D. Johnson, "Computer-assisted detection of colonic polyps with CT colonography using neural networks and binary classification trees," *Med. Phys.*, vol. 30, pp. 52–60, 2003.
48. J. Näppi, H. Frimmel, A. H. Dachman, and H. Yoshida, "A new high-performance CAD scheme for the detection of polyps in CT colonography," in *Med Imaging 2004: Image Processing*, San Diego, CA, February 14, pp. 839–848, 2004.

49. A. K. Jerebko, J. D. Malley, M. Franaszek, and R. M. Summers, "Multiple neural network classification scheme for detection of colonic polyps in CT colonography Databases," *Acad. Radiol.*, vol. 10, pp. 154–160, 2003.
50. A. K. Jerebko, J. D. Malley, M. Franaszek, and R. M. Summers, "Support vector machines committee classification method for computer-aided polyp detection in CT colonography," *Acad. Radiol.*, vol. 12, pp. 479–86, 2005.
51. D. P. Zhu, R. W. Conners, D. L. Schmoldt, and P. A. Araman, "A prototype vision system for analyzing CT imagery of hardwood logs," *IEEE Trans. Syst. Man Cyb. B* , vol. 26, pp. 522–532, 1996.
52. Y. Shao, C. H. Chang, "Bayesian separation with sparsity promotion in perceptual wavelet domain for speech enhancement and hybrid speech recognition," *IEEE Trans. Syst. Man Cyb. A*, vol.41, pp. 284–293, 2007.
53. V. F. van Ravesteijn, C. van Wijk, F. M. Vos, R. Truyen, J. F. Peters, J. Stoker, and L. J. van Vliet, "Computer-aided detection of polyps in CT colonography using logistic regression," *IEEE Trans. Med. Imaging*, vol. 29, pp. 120–131, 2010.
54. C. L. Cai, J. G. Lee, M. E. Zalis, and H. Yoshida, "Mosaic decomposition: An electronic cleansing method for inhomogeneously tagged regions in noncathartic CT colonography," *IEEE Trans. Med. Imaging*, vol.30, pp. 559–570, 2011.
55. X Geng, D.C. Zhan and Z.H. Zhou, "Supervised nonlinear dimensionality reduction for visualization and classification," *IEEE Trans. Syst. Man Cyb. B*, vol.35, pp. 1098–1107. 2005.
56. C. F. Juang, S. H. Chiu, and S. J. Shiu, "Fuzzy system learned through fuzzy clustering and support vector machine for human skin color segmentation," *IEEE Trans. Syst. Man Cyb. A*, vol.37, pp. 1077–1087, 2007.
57. P. Chung, C. Chang, W. C. Chu, and H. C. Lin, "Reconstruction of medical images under different image acquisition angles," *IEEE Trans. Syst. Man Cyb. B*, vol. 33, pp. 503–509, 2003.
58. T. Fawcett, "An introduction to ROC analysis," *Pattern Recogn. Lett.*, vol. 27, no. 8, 2006.
59. R. O. Duda, P. E. Hart, and D. G. Stork, *Pattern classification*. Hoboken, NJ: John Wiley & Sons, 2001.
60. M. F. McNitt-Gray, H. K. Huang, and J. W. Sayre, "Feature selection in the pattern classification problem of digital chest radiograph segmentation," *IEEE Trans. Med. Imaging*, vol. 14, pp. 537–547,1995.
61. P. J. Pickhardt, J. R. Choi, I. Hwang, J. A. Butler, M. L. Puckett, H. A. Hildebrandt, R. K. Wong, P. A. Nugent, P. A. Mysliwiec and W. R. Schindler, "Computed tomographic virtual colonoscopy to screen for colorectal neoplasia in asymptomatic adults," *N. Engl. J. Med.*, vol.349, pp. 2191–2200, 2003.
62. D. C. Rockey, E. Paulson, D. Niedzwiecki, W. Davis, H. B. Bosworth, L. Sanders, J. Yee, J. Henderson, P. Hatten, S. Burdick, A. Sanyal, D. T. Rubin, M. Sterling, G. Akerkar, M. S. Bhutani, K. Binmoeller, J. Garvie, E. J. Bini, K. McQuaid, W. L. Foster, W. M. Thompson, A. Dachman, and R. Halvorsen,"Analysis of air contrast barium enema, computed tomographic colonography, and colonoscopy: Prospective comparison," *Lancet*, vol. 365, pp. 305–311, 2005.
63. D. Regge, C. Laudi, G. Galatola, P. Della Monica, L. Bonelli, G. Angelelli, R. Asnaghi, B. Barbaro, C. Bartolozzi, D. Bielen, L. Boni, C. Borghi, P. Bruzzi, M. C. Cassinis, M. Galia, T. M. Gallo, A. Grasso, C. Hassan, A. Laghi, M. C. Martina, E. Neri, C. Senore, G. Simonetti, S. Venturini, and G. Gandini, "Diagnostic accuracy of computed tomographic

colonography for the detection of advanced neoplasia in individuals at increased risk of colorectal cancer," *JAMA*, vol. 301, no. 23, pp. 2453–2461, 2009.

64. Department Cluster Dogwood (online). http://computer-science.egr.vcu.edu/undergraduate/resourcescurrent/computing-facilities/. Accessed 20 November 2014.
65. M. Awad, Y. Motai, J. Näppi, and H. Yoshida, "A clinical decision support framework for incremental polyps classification in virtual colonoscopy," *Algorithms*, vol. 3, pp.1–20, 2010.
66. X. Jiang, R. Snapp, Y. Motai, and X. Zhu, "Accelerated kernel feature analysis," in *Proc. IEEE Computer Society Conference on Computer Vision and Pattern Recognition*, New York, NY, June 17–22, pp. 109–116, 2006.
67. B. Schökopf, and A. J. Smola. "Learning with kernels: support vector machines, regularization, optimization, and beyond," *Adaptive computation and machine learning*. Cambridge, MA: MIT Press, 2002.
68. L. Jayawardhana, Y. Motai, and A. Docef, "Computer-aided detection of polyps in CT colonography: On-line versus off-line accelerated kernel feature analysis," Special Issue on Processing and Analysis of High-Dimensional Masses of Image and Signal Data,*Signal Process.*, pp.1–12, 2009.
69. Amazon Elastic Compute Cloud (online). http://aws.amazon.com/ec2/. Accessed 20 November 2014.
70. H Yoshida, J Näppi, and W Cai, "Computer-aided detection of polyps in CT colonography: Performance evaluation in comparison with human readers based on large multicenter clinical trial cases," in *Proc 6th IEEE Int. Symp. Biomedical Imaging*, Boston, MA, June 28–July 1, pp. 919–922, 2009.
71. H. Xiong, M. N. S. Swamy, and M. O. Ahmad, "Optimizing the data-dependent kernel in the empirical feature space," *IEEE Trans. Neural Netw.*, vol. 16, pp. 460–474, 2005.
72. N. Cristianini, J. Kandola, A. Elisseeff, and J. Shawe-Taylor, "On kernel target alignment," *Proc. Neural Inform. Process. Syst.*, pp. 367–373, 2001.
73. Berkeley Open Infrastructure for Network Computing (online). http://boinc.berkeley.edu/. Accessed 20 November 2014.
74. Microsoft Azure (online). https://azure.microsoft.com. Accessed 20 November 2014.
75. IBM cloud services (online). http://www.ibm.com/cloud-computing/us/en/. Accessed 20 November 2014.
76. H. Yoshida and J. Nappi, "Three-dimensional computer-aided diagnosis scheme for detection of colonic polyps," *IEEE Trans. Med. Imaging*, vol. 20, pp. 1261–1274, 2001.
77. J. Näppi and H. Yoshida, "Fully automated three-dimensional detection of polyps in fecal-tagging CT colonography," *Acad. Radiol.*, vol. 14, pp. 287–300, 2007.
78. MathWorks (online). http://www.mathworks.com/products/parallel-computing/.Accessed 20 November 2014.

6

PREDICTIVE KERNEL ANALYSIS

6.1 INTRODUCTION

Anomaly detection has long been an obstacle in the detection and treatment of colorectal cancer. The current state-of-the-art diagnosis techniques still fail to identify the important transitional cases over the historical aspects of patient datasets. Anomaly detection in nonstationary medical data is critical for detecting and adapting to changes in the course of the diagnosis and treatment of ongoing cancer [1]. In computed tomography (CT) colonoscopy (also known as virtual colonoscopy [2]), for example, it has been desired for the improvement of automated detection performance of computer-aided diagnosis schemes [3–6] in differentiating polyps from false positives on the basis of longitudinal data. The main problem pertaining to nonstationary CT colonography datasets is how to detect and predict an anomaly or change in the data streams over time. Thus, the purpose of this study is to detect anomaly cases on the basis of a clinical longitudinal CT colonography datasets.

The data acquired "over a long period of time" are sometimes highly diverse and suffer from numerical errors. Therefore, obtaining a clear distinction between "anomaly" and "normal" large nonstationary datasets is a highly challenging task. The specific problems to address in this study on anomaly detection are (i) to introduce the criteria of the anomaly degree and (ii) to predict the consistent progress of the anomaly degree in the dynamic nonstationary environment. Nonstationary cancer data in real-world applications are highly nonlinear and unbalanced, underlying distribution changes over time.

To address the problems of nonstationary, nonlinear, and unbalanced datasets, various state-of-the-art methods on pattern classification [7–12] have been developed. Indeed, detecting and adapting classifiers to changes in underlying data distributions

Data-Variant Kernel Analysis, First Edition. Yuichi Motai.

is an active area of research [10]. For example, Gönen et al. [13] proposed a multiple kernel learning (MKL) method, similar to batch learning, in which the input data was divided into a few groups of similar sizes, and kernel principal component analysis (KPCA) was applied to each group [14, 15]. The application of the online concept to principal component analysis was mostly referred to incremental principal component analysis [16–20] with computational effectiveness in many image processing applications and pattern classification systems [21–24]. The effective application of online learning to nonlinear space was already undertaken using kernel-based methods [25–28]. The general problem addressed by this study is the inability of these state-of-the-art methods to accommodate a stream or batches of data whose underlying distribution changes over time.

In this study, we propose a new nonstationary learning technique called longitudinal analysis of composite kernels (LACK), which substantially extends the following stationary learning techniques: accelerated kernel feature analysis (AKFA) [29] and principal composite kernel feature analysis (PC-KFA) [30]. The proposed LACK algorithm is applied to the real-world texture-based features of polyp candidates generated by a shape-based computer-aided detection (CAD) scheme in an effort to validate its ability to detect and predict anomalies. Constructing a composite kernel from data-dependent kernels [30, 31] is a solution for the anomaly data problem. The composite kernel will modify itself to make kernel choice more optimized. To compromise the weakness of traditional composite kernels, we have developed nonstationary data associations using LACK to correspond with anomaly datasets of nonstationary cases. The contribution of this study is the design and development of a new LACK method that yields high classification and prediction accuracy in cancer staging. The proposed LACK method was evaluated on the basis of the texture features of polyps detected by a CAD scheme from the clinical CT colonography database to determine its performance in the detection and prediction of anomalies. We have shown that the overall computing cost does not substantially change due to the light module of nonstationary prediction.

The remainder of this chapter is organized as follows. Section 6.2 provides an introduction to kernels and a brief review of the existing kernel-based feature extraction methods KPCA and AKFA. In Section 6.3, we present the proposed composite data-dependent kernel feature analysis for the detection of polyp candidates in the stationary data. In Section 6.4, we extend this stationary kernel method to accommodate normal and anomaly cases into nonstationary kernel method. In Section 6.5, we describe how the LACK method detects and predicts the status, "normal" or "anomaly," from the incoming dataset. The experimental results of classification are drawn in Section 6.6, and the results of longitudinal prediction are drawn in Section 6.7. Finally, Section 6.8 presents the conclusions.

6.2 KERNEL BASICS

For efficient feature analysis, extraction of the salient features of polyps is essential because of the size and the three-dimensional nature of the polyp datasets. Moreover, the distribution of the image features of polyps is expected to be nonlinear.

The problem is how to select such an ideal nonlinear positive-definite kernel operator: $k : R^d \times R^d \to R$ of an integral operator. The kernel K is a positive semidefinite matrix that computes the inner product between any two finite sequences of inputs $x := \{x_j : j \in N_n\}$ and is defined as: $K := (K(x_i, x_j) : i, j \in N_n) = (\varphi(x_i) \cdot \varphi(x_j))$. Some of the commonly used kernels are [32]

The linear kernel:

$$K(x, x_i) = x^T x_i \tag{6.1}$$

The polynomial kernel:

$$K(x, x_i) = (x^T x_i + \mathit{offset})^d \tag{6.2}$$

The Gaussian radial basis function (RBF) kernel:

$$K(x, x_i) = \exp(-\|x - x_i\|^2 / 2\sigma^2) \tag{6.3}$$

The Laplace RBF kernel:

$$K(x, x_i) = \exp(-\sigma \|x - x_i\|) \tag{6.4}$$

The sigmoid kernel:

$$K(x, x_i) = \tanh(\beta_0 x^T x_i + \beta_1) \tag{6.5}$$

The analysis of variance (ANOVA) RB kernel:

$$K(x, x_i) = \sum_{k=1}^{n} \exp\left(-\sigma(x^k - x_i^k)^2\right)^d \tag{6.6}$$

Kernel selection is heavily dependent on the data specifics. For instance, the linear kernel is important in large sparse data vectors and it implements the simplest of all kernels, whereas the Gaussian and Laplace RBFs are general purpose kernels used when prior knowledge about data is not available. The Gaussian kernel avoids the sparse distribution caused by the high-degree polynomial kernel in large feature space. The polynomial kernel is widely applied in image processing, while the ANOVA RB is usually reserved for regression tasks. For these reasons, in our study, we are using the kernels in Equations 6.1–6.4 to form our data-dependent composite kernel. A more thorough justification can be found in [32–36].

Our new kernel feature analysis for CT colonographic images is based on the following two methods: KPCA [25–28] and AKFA [29]. We proceed with a brief review of these existing methods.

6.2.1 KPCA and AKFA

KPCA uses a Mercer kernel [25, 37, 38] to perform a linear principal component analysis of the transformed image. Without loss of generality, we assume that the image of the data has been centered so that its scatter matrix in H is given by

$S = \sum_{i=1}^{n}(\Phi(x_i)\Phi(x_i)^T$. The eigenvalues λ_j and eigenvectors e_j are obtained by solving the following equation:

$$\lambda_j e_j = Se_j = \sum_{i=1}^{n} \Phi(x_i)\Phi(x_i)^T e_j = \sum_{i=1}^{n} \langle e_j, \Phi(x_i)\rangle \Phi(x_i). \tag{6.7}$$

If K is an $n \times n$ Gram matrix, with $k_{ij} = \langle \Phi(x_i), \ \Phi(x_j)\rangle$ and $a_j = [a_{j1}a_{j2} \cdots a_{jn}]^T$ associated with eigenvalues $\lambda_{j,}$ then the dual eigenvalue problem equivalent to Equation 6.7 can be expressed as follows $\lambda_j a_j = Ka_{j.}$ Let us keep the l eigenvectors associated with the l largest eigenvalues, we can reconstruct data in the mapped space as follows: $\Phi_i^{'} = \sum_{j=1}^{\ell} \langle \Phi_i, e_j\rangle e_j = \sum_{j=1}^{\ell} \beta_{ji} e_j$, where $\beta_{ji} = \left\langle \Phi_i, \sum_{k=1}^{n} a_{jk}\Phi_k \right\rangle = \sum_{k=1}^{n} a_{jk}k_{ik}$.

KPCA can now be summarized as follows:

Step 1. Calculate the Gram matrix using kernels from Equations 6.1 to 6.4, which contains the inner products between pairs of image vectors.

Step 2. Use $\lambda_j a_j = Ka_j$ to get the coefficient vectors a_j for $j = 1, \ldots, n$.

Step 3. The projection of a testing point $x \in R^d$ along the jth eigenvector is

$$\langle e_j, \ \Phi(x)\rangle = \sum_{i=1}^{n} a_{ji} \langle \Phi(x_i), \Phi(x)\rangle = \sum_{i=1}^{n} a_{ji} k(x, \ x_i) \tag{6.8}$$

The above equation implicitly contains an eigenvalue problem of rank n, so the computational complexity of KPCA is $O(n^3)$. In addition, each resulting eigenvector is represented as a linear combination of n terms. Thus, all data contained in X_n must be retained, which is computationally cumbersome and unacceptable for nonstationary training.

AKFA can be summarized as follows [29]:

Step 1. Compute the $n \times n$ Gram matrix $k_{ij} = k\langle x_i, \ x_j\rangle$, where n is the number of input vectors. Let l denote the number of features to be extracted. Initialize the $\ell \times \ell$ coefficient matrix $\mathbf{C}$ to 0, and $idx(\cdot)$ as an empty list, which will ultimately store the indices of the selected image vectors, and $\mathbf{C}_{(i-1)}$ is an upper-triangle coefficient matrix. Let us define, $\Phi^{i}_{idx(i)} = \Phi_{idx(i)} - \sum_{t=1}^{i-1} \langle \Phi_{idx(i)}, v_t\rangle v_t$. Initialize the threshold value $\delta = 0$ for the reconstruction error.

Step 2. For $i = 1$ to ℓ, using the ith updated $\mathbf{K^i}$ matrix, extract the ith feature. If $K^i_{jj} < 0$, the predetermined $\delta > 0$. It is a threshold that determines the number of features we selected. Then, discard jth column and jth row vector without calculating the projection variance. Use $idx(i)$ to store the index.

Step 3. Update the coefficient matrix by using $\mathbf{C}_{i,i} = 1/\sqrt{k^i_{idx(i),idx(i)}}$. and $\mathbf{C}_{1:(i-1),i} = -\mathbf{C}_{i,i}\mathbf{C}_{(\mathbf{i-1})}\mathbf{C}^T_{(\mathbf{i-1})}\mathbf{K}_{\mathbf{idx(i)}}$.

Step 4. Obtain $\mathbf{K}^{i+1}$, neglect all rows and columns containing diagonal elements less than δ, return to Step 2.

The AKFA algorithm also contains an eigenvalue problem of rank n, so the computational complexity of AKFA in Step 1 requires $O(n^2)$ operations, Step 2 is $O(\ell^2)$. Step 3 requires $O(n^2)$. The total computational complexity is increased to $O(\ell n^2)$ when no data is being truncated. Once we have our Gram matrix ready, we can apply any one of these algorithms to our dataset to obtain a higher dimensional feature space. This idea is discussed in the following sections.

6.3 STATIONARY DATA TRAINING

The success of the KPCA and AKFA depends heavily on the kernel used for constructing the Gram matrix. Therefore, the key learning element is to adapt a problem-specific kernel function [37, 39, 40]. We will briefly address a data-dependent approach for stationary learning in Sections 6.3.1 and 6.3.2.

6.3.1 Kernel Selection

We exploit the idea of the data-dependent kernel that is presented in [12] to select the most appropriate kernels for a given dataset. Let $\{x_i, y_i\}(i = 1, 2, \dots, n)$ be n d-dimensional training samples of the given data, where $y_i = \{+1, -1\}$ represents the class labels of the samples. The data-dependent composite kernel k_r for $r = 1, 2, 3, 4$ can be formulated as

$$k_r(x_i, x_j) = q_r(xi)q_r(xj)p_r(xi, xj) \tag{6.9}$$

where, $x \in R^d$, $p_r(xi, xj)$ is one kernel among five chosen kernels (Eqs. 6.1–6.4), and $q(\cdot)$ is the factor function $q_r(x_i) = \alpha_{r0} + \sum_{m=1}^{n} \alpha_{rm} k_0(x_i, x_m)$, where $k_0(x_i, x_m) = \exp(-\|x_i - x_m\|^2 / 2\sigma^2)$, and α_{rm} are the combination coefficient. Let us denote the vectors $\{a_{r0}\ a_{r1}, a_{r2}, \dots, a_{rm}\}$ and $\{q_r(x_1), q_r(x_2), \dots, q_r(x_n)^T\}$ by α_r and q_r, respectively.

We have $q_r = K_0 \alpha_r$ where K_0 is a $n \times (n+1)$ matrix given as

$$K_0 = \begin{bmatrix} 1 & k_0(x_1, x_1) & \cdots & k_0(x_1, x_n) \\ 1 & k_0(x_2, x_1) & \cdots & k_0(x_2, x_n) \\ \vdots & \vdots & \cdots & \vdots \\ 1 & k_0(x_n, x_1) & \cdots & k_0(x_n, x_n) \end{bmatrix}.$$

Let the data-dependent kernel matrix K_r [40] correspond to $k(x_i, x_j)$:

$$K_r = [q_r(x_i)q_r(x_j)p_r(x_i, xj)]_{n \times n} \tag{6.10}$$

Defining Q_i as the diagonal matrix of elements $\{q_r(x_1), q_r(x_2), \dots, q_r(x_n)\}$, and P_r as the elements of $p_r(x_i, x_j)$, we can express Equation 6.10 as

$$K_r = Q_r P_r Q_r \tag{6.11}$$

The optimization of the data-dependent kernel [31] in Equation 6.9 consists of choosing the value of combination coefficient vector α_r so that the class separability of the training data in the mapped feature space is maximized. For this purpose, we use the Fisher scalar as the objective function of our kernel optimization. The Fisher scalar is used to measure the class separability J of the training data formulated as

$$J = \frac{\left(\mathrm{tr}\left(\sum_r S_{\mathrm{br}}\right)\right)}{\left(\mathrm{tr}\left(\sum_r S_{\mathrm{wr}}\right)\right)} \tag{6.12}$$

where S_{br} represents "between-class scatter matrices" and S_{wr} represents "within-class scatter matrices." Suppose that the training data are grouped according to their class labels, that is, the first n_1 data belongs to one class, and the remaining n_2 data belongs to the other classes ($n_1 + n_2 = n$). Then, the basic kernel matrix P_r can be partitioned as $P_r = [P^r_{11}, P^r_{12}; P^r_{21}, P^r_{22}]$, where $r = 1, 2, 3, 4$ represents the kernels from Equations 6.1 to 6.4, and the size of the submatrices P^r_{11}, P^r_{12}, P^r_{21}, P^r_{22} are $n_1 \times n_1, n_1 \times n_2, n_2 \times n_1, n_2 \times n_2$, respectively. According to [41], the class separability in Equation 6.12 can be expressed as

$$J(\alpha_r) = \frac{\alpha_r^T M_{0r} \alpha_r}{\alpha_r{}^T N_{0r} \alpha_r} \tag{6.13}$$

where $M_{0r} = K_0^T B_{0r} K_0$, and $N_{0r} = K_0^T W_{0r} K_0$ $B_{0r} = \begin{pmatrix} \frac{1}{n_1} P^r_{11} & 0 \\ 0 & \frac{1}{n_2} P^r_{22} \end{pmatrix} - \frac{1}{n} P_r$, $W_{0r} = \mathrm{diag}\,(p^r_{11}, p^r_{22}, \dots, p^r_{nn}) - \begin{pmatrix} \frac{1}{n_1} P^r_{11} & 0 \\ 0 & \frac{1}{n_2} P^r_{22} \end{pmatrix}$.

The vectors are denoted $\{\alpha_{r0}, \alpha_{r1}, \alpha_{r2}, \ \dots, \ \alpha_{rn}\}$ and $\{q_r(x_1), q_r(x_2), \dots, q_r(x_n)\}^T$ by α_r and q_r.

To maximize $J(\alpha_r)$ in Equation 6.13, the standard gradient approach is followed. If the matrix N_{0i} is nonsingular, the optimal α_r that maximizes the $J(\alpha_r)$ is the eigenvector corresponding to the maximum eigenvalue of the system,

$$M_{0r} \alpha_r = \lambda_r N_{0r} \alpha_r \tag{6.14}$$

The criterion for selecting the best kernel function is finding the kernel that produces the largest eigenvalue from Equation 6.14, that is, $\lambda_r^* = \arg\max_{\lambda}(N_r{}^{-1} M_r)$. The maximum eigenvector α corresponding to the maximum eigenvalue will result in the optimum solution. Once we derive the eigenvectors, we now proceed to construct q_r ($q_r = K_0 \alpha_r(n)$) and Q_r to find the corresponding Gram matrices of these kernels. Then, we can find the optimum kernels for the given dataset. To do this, we have to arrange the eigenvalues (which determined the combination coefficients) for all kernel functions in descending order. The first p kernels corresponding to the first p eigenvalues will be used in the construction of a composite kernel that is expected to yield optimum classification accuracy. This method is described in the following subsection.

6.3.2 Composite Kernel: Kernel Combinatory Optimization

A composite kernel function is defined as the weighted sum of the set of different optimized kernel functions. To obtain the optimum classification accuracy, we define the composite kernel as

$$K_{\text{comp}}(\rho) = \sum_{i=1}^{p} \rho_i Q_i P_i Q_i \tag{6.15}$$

where the value of the composite coefficient ρ_i is a scalar value and p is the number of kernels we intend to combine. According to [42], this composite kernel matrix $K_{\text{comp}}(\rho)$ satisfies Mercer's condition.

Now, the problem is how to determine this composite coefficient $\hat{\rho}$ ($\hat{\rho} = [\rho_1, \rho_2, \dots, \rho_p]$) such that the classification performance is optimized. To this end, we adapted the concept of "kernel alignment" to determine the best $\hat{\rho}$ that gives us optimum performance. The "Alignment" measure was introduced by Cristianini et al. [43] to measure the adaptability of a kernel to the target data and provide a practical objective for kernel optimization. The empirical alignment between kernel K and kernel yy^T is

$$A(k, yy^T) = \frac{\langle K, yy^T \rangle_F}{\|K\|_F \|yy^T\|_F} = \frac{y^T K y}{n\|K\|_F} \tag{6.16}$$

It has been shown that if a kernel is well aligned with the target information, there exists a separation of the data with a low bound on the generalization error [33]. Let us consider the combination of kernel functions corresponding to Equation 6.16 as $k(\rho) = \sum_{i=1}^{p} \rho_i k_i$ where the kernels k_i, $i = 1, 2, \dots, p$ are known in advance. Our purpose is to tune ρ to maximize $A(\rho,\ k,\ yy')$ the empirical alignment between $k(\rho)$ and the target vector y. Hence,

$$\begin{aligned}
\hat{\rho} &= \underset{\rho}{\operatorname{argmax}}(A(\rho, k, yy^T)) \\
&= \underset{\rho}{\operatorname{argmax}} \frac{\left(\sum_i \rho_i u_i\right)^2}{\left(n^2 \sum_{i,j} \rho_i \rho_j v_{ij}\right)} = \underset{\rho}{\operatorname{argmax}} \left(\frac{1}{n^2} \cdot \frac{\rho^T U \rho}{\rho^T V \rho}\right)
\end{aligned} \tag{6.17}$$

where $u_i = \sqrt{\langle K_i, yy^T \rangle}$, $v_{ij} = \sqrt{\langle K_i, K_j \rangle}$, $U_{ij} = u_i u_j$, $V_{ij} = v_i v_j$, $\rho = (\sqrt{\rho_1}, \sqrt{\rho_2}, \dots, \sqrt{\rho_p})$.

Let the generalized Raleigh coefficient be given as $J(\rho) = \rho^T U \rho / \rho^T V \rho$.

We can obtain $\hat{\rho}$ by solving the generalized eigenvalue problem $U\rho = \delta V \rho$ where δ denotes the eigenvalues. We find this optimum composite coefficient $\hat{\rho}$, which will be the eigenvector corresponding to maximum eigenvalue δ. In this way, we find the "kernel Gram matrix" for the stationary data.

6.4 LONGITUDINAL NONSTATIONARY DATA WITH ANOMALY/NORMAL DETECTION

The new criteria in training the nonstationary longitudinal data have been desired [1]. Among the incoming nonstationary datasets, there is a chance that the data may be misinterpreted as either "normal" or "anomalous." The key idea of the classification of normal/anomaly is to allow the feature space to be updated efficiently as the training proceeds with more data being fed into the nonstationary algorithm. The following subsections will explore updating the Gram matrix based on nonstationary longitudinal data (Section 6.4.1) and composite kernel for nonstationary data (Section 6.4.2).

6.4.1 *Updating the Gram Matrix Based on Nonstationary Longitudinal Data*

We use "class separability" as a measure to identify whether the nonstationary data is an anomaly or normal. As our terminal definition, we call the data normal when it improves the class separability. Conversely, the data is anomalous when the class is degenerated with the nonstationary data. Let us introduce a variable ξ., which is the ratio of the class separability of the composite (both stationary and nonstationary) data and the stationary data. It can be expressed as

$$\xi = \frac{J'_*(\alpha'_r)}{J_*(\alpha_r)} \tag{6.18}$$

where $J'_*(\alpha'_r)$ denotes the class separability yielded by the most dominant kernel (which is chosen from the four different kernels) for the composite data (i.e., new incoming data and the previous stationary data), and $J_*(\alpha_r)$ is the class separability yielded by the most dominant kernel for stationary data. Because of the properties of class separability, Equation 6.18 can be rewritten as $\xi = \lambda'_* / \lambda_*$, where λ'_* correspond to the most dominant eigenvalue of composite data (both stationary and nonstationary), and λ_* is the most dominant eigenvalue of the four different kernels for the stationary data. If ξ is less than a threshold value η, then the incoming nonstationary data is anomaly; otherwise, it is normal.

Normal Nonstationary Data In the case of normal data, we do not update the Gram matrix. Instead, we discard all the data that is normal. Hence, the updated Gram matrix (which is exactly the same as Eq. 6.14)) can be given as $K_r^{n\prime} = Q_r P_r Q_r$.

Anomaly Nonstationary Data If the new batch of incoming data contains an anomaly, the feature space is updated. But this update can be either nonincremental if the nonstationary data is a minor anomaly (Case 1) or nonincremental if the nonstationary data is a significant anomaly (Case 2) by introducing the following criterion called "residue factor."

The class separability of the most dominant kernel for the new data is directly dependent on both the maximum combination coefficient of four different kernels as well as the maximum eigenvalue λ'_*. Let us denote α_* as the mean of combination coefficients prospectively, and α'_* for the most dominant kernel among the four kernels available. Using these values, we introduce new criteria to determine whether the update should be incremental or nonincremental by evaluating disparities between composite (nonstationary and stationary) and offline eigen representation. The residue factor "rf" is defined as

$$rf = (\alpha'_* - \alpha_*) \cdot \frac{\lambda'_*}{\lambda_*} \tag{6.19}$$

The α_* is the weight in kernel defined in Equation 6.18, and its value represents the power of the corresponding element in the kernel. λ'_* is the maximal eigenvalue of composite kernel to present adaptivity of kernel for the dataset. When residue factor is greater than one, similarity between old and new eigenvectors is very small, so Equation 6.19 will be used as a criterion for incremental or nonincremental update.

Case 1: Minor Anomaly If the residue factor rf (Eq. 6.19) is less than 1, then the update is nonincremental. Hence, the dimensions of the Gram matrix remain constant. We define small elements in rf as trivial elements and corresponding rows and columns in the Gram matrix as trivial rows and columns. We replace the trivial rows and columns of the previous Gram matrix with those of the new data by calculating the minimum difference vectors minΔ as the trivial one. We compare and replace the combination coefficient values of stationary data with the combination coefficient of the new incoming data as shown in Fig. 6.1. This process should be repeated for all the kernel matrices.

The input matrices P' and Q' are updated by removing the row and column corresponding to the index of $\alpha_{r\min\Delta}$ and replacing it with the row and column corresponding to the index α'_r. Then, the new input matrix $P_r^{n'}$ can be written as

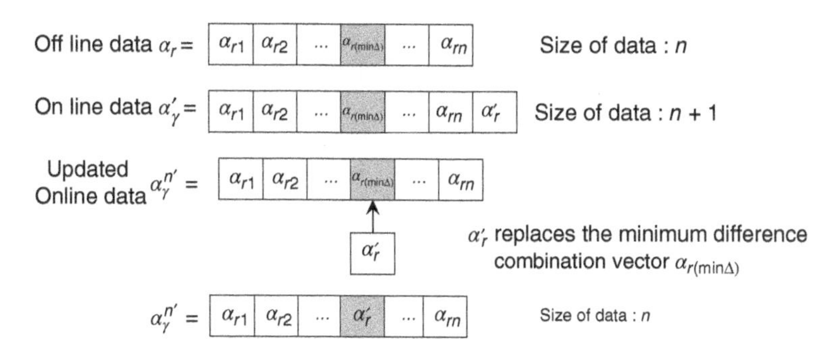

Figure 6.1 The illustration of a nonincremental update of a combination vector. The trivial rows and columns of the previous Gram matrix with those of the new data are replaced as follows $\alpha_{r(\min\Delta)} \leftarrow \alpha'_r$.

$P_r^{n'} = [P_r'] - [P_{r(r\min\Delta,1:(n+1))}] - [P_{r(1:n,r\min\Delta)}]$. This can be shown as follows:

$$P_r^{n'} = \begin{bmatrix} p_{r(1,1)} & \cdot & p_{r(1,\min\Delta)} & \cdot & p_{r(1,(n+1))} \\ \cdot & \cdot & \cdot & \cdot & \cdot \\ [p_{r(\min\Delta,1)} & \cdot & p_{r(\min\Delta,\min\Delta)} & \cdot & p_{r(\min\Delta,n+1)}] \\ \cdot & \cdot & \cdot & \cdot & \cdot \\ p_{r((n+1),1)} & \cdot & p_{r((n+1),\min\Delta)} & \cdot & p_{r((n+1),(n+1))} \end{bmatrix}$$
$$= \begin{bmatrix} p_{r(1,1)} & \cdots & p_{r(1,(n+1))} \\ \vdots & \cdots & \vdots \\ p_{r(n,1)} & \ddots \cdots & p_{r(n,(n+1))} \end{bmatrix}_{n\times n(n+1)\times(n+1)}$$

After we compute $\alpha_r^{n'}$, we can compute $Q^{n'}$ as $Q_r{}^{n\prime} = \text{diag}(\alpha^{n\prime})$. Hence, in the nonincremental update of minor anomaly data, the Gram matrix can be given as $K_r^{n'} = Q_r^{n'} P_r^{n'} Q_r^{n'}$.

Case 2: Significant Anomaly If the residue factor is greater than one, it means the similarity between the previous eigenvectors and the new eigenvectors is very small. So it is very important for us to retain this new data, which is highly anomalous, for efficient classification. Therefore, instead of replacing the trivial rows and columns of the previous data, we simply retain it and, thus, the size of the Gram matrix is incremented by the size of the new data. In this case, as we have already calculated the new combination coefficient α_r', input matrix P' and Q' as same as in Section 6.3. Then, the kernel Gram matrix can be given as $K_r^{n'} = Q_r' P_r' Q_r'$.

Once we have our kernel Gram matrices, we can now determine the composite kernel that will give us the optimum classification accuracy when the existing stationary data is incorporated with new nonstationary data, shown in the following subsection.

6.4.2 Composite Kernel for Nonstationary Data

As we have computed the Gram matrices for the (p)-most dominant kernels for the composite data (stationary plus nonstationary data), we now combine them to yield the best classification performance. As defined in Section 6.3.1, we can write the composite kernel for the new composite data as $K_{\text{comp}}^{n'}(\rho) = \sum_{i=1}^{p} \rho_i' Q_i^{n'} P_i^{n'} Q_i^{n'}$, where the composite coefficient set $\hat{\rho}'$ is the collection of combination coefficients ρ_i', and p is the number of kernels we intend to combine. We use the same alignment factor method to determine the optimum composite coefficients that will yield the best composite kernel. Let our new label vector be $y^{n\prime}$. Similar to stationary Section 6.3.2, we may apply the same technique to update the Gram matrix as shown in Fig. 6.2. The new alignment factor can be written from Equation 6.16 $A\prime(k^{n\prime}, y^{n\prime}(y^{n\prime})^T) = ((y^{n\prime})^T K^{n\prime} y^{n\prime})/(n' \| K^{n\prime} \|_F)$.

We can determine the composite coefficient ρ' that will maximize the alignment factor (Eqs. 6.17–6.19) as follows:

$$\hat{\rho}' = \arg\max_{\rho'}(A'(\rho', k^{n'}, y^{n'}(y^{n'})^T)) = \arg\max_{\rho'} \left(\frac{1}{n'^2} \frac{(\rho_i')^T U^{n'} \rho_i'}{(\rho_i')^T V^{n'} \rho_i'} \right) \tag{6.20}$$

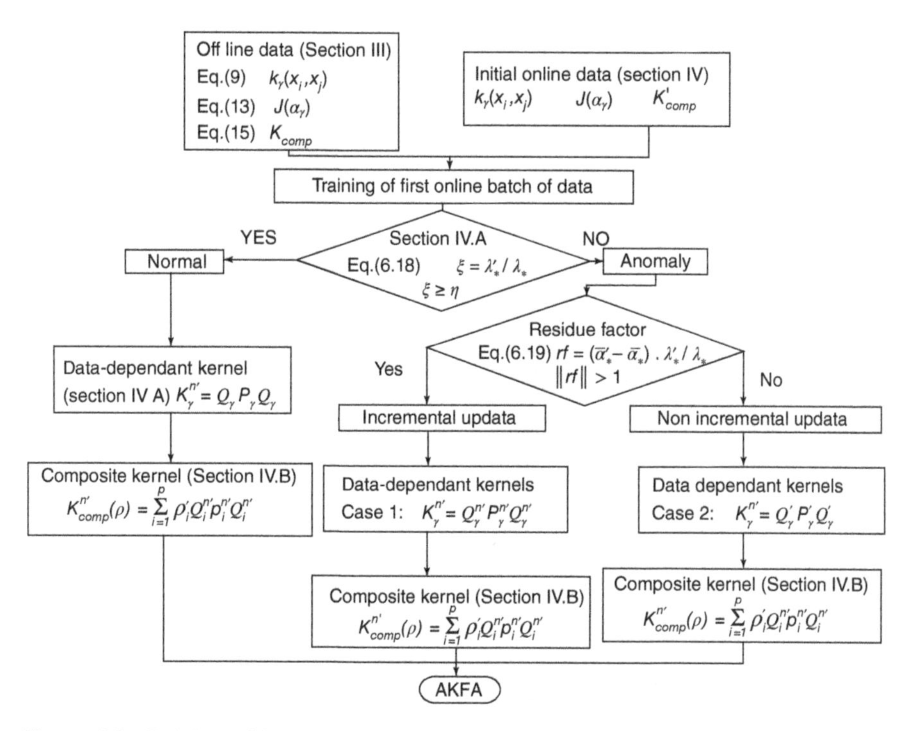

Figure 6.2 Training of first nonstationary batch of data using two cases to iteratively update composite kernels. Anomaly or normal detection is conducted using the proposed criterion.

where $u_i^{n'} = \langle K_i^{n'}, y^{n'}(y^{n'})^T \rangle$, $U_{ij}^{n'} = u_i^{n'} u_j^{n'}$, $V_{ij}^{n'} = v_{ij}^{n'} = \langle K_i^{n'}, K_j^{n'} \rangle$. Let the new generalized Raleigh coefficient $J(\rho') = ((\rho')^T U^{n'} \rho')/((\rho')^T V^{n'} \rho')$. We choose a composite coefficient vector that is associated with the maximum eigenvalue. Once we know the eigenvectors, that is, combination coefficients of the composite data (both stationary and nonstationary), we can compute q'_r and hence Q'_r to find out the four Gram matrices corresponding to the four different kernels by using $K_r^{n'} = Q'_r P'_r Q'_r$. Now, the first p kernels corresponding to the first p eigenvalues (arranged in the descending order) will be used in the construction of a composite kernel that will yield optimum classification accuracy. In the following section, we see how to handle the nonstationary data over a long period of time.

6.5 LONGITUDINAL SEQUENTIAL TRAJECTORIES FOR ANOMALY DETECTION AND PREDICTION

When the longitudinal trajectory is considered, the long-term sequential change is evaluated according to the criteria with the anomaly degree. The nonstationary data is suffering from the large memory size and is required to accommodate a stream or

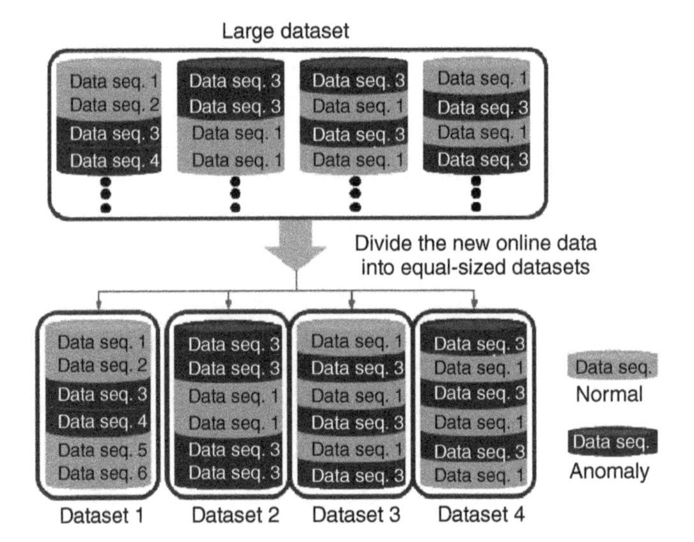

Figure 6.3 A conceptual illustration of the division of a large nonstationary dataset into several small subsets of equal size. Large dataset is clustered into small subsets of datasets, for example, 1-2-3-4, so make the feasible size of the time sequence.

batches of data whose underlying distribution changes over time. The long-term nonstationary data is divided into smaller sizes in the half amount iteratively, and each of which is considered one at a time. We may then treat small datasets as anomaly or normal data by checking the class separability of each dataset. Let us consider a small subset of data to find the appropriate window size for the incoming data and determine whether it is normal or anomaly data using Equation 6.19. The following Fig. 6.3 illustrates the segmentation of the incoming nonstationary data into small equal subsets l. All new sets of data have to be processed sequentially but not simultaneously.

6.5.1 Anomaly Detection of Nonstationary Small Chunks Datasets

In order to track whether we are correctly training the algorithm, we make use of the Alignment factor introduced in Section 6.5.2. Let us denote ${}^{t+1}(K^{n'}_{comp})_t$ as the update of the Gram matrix from time t to $t+1$, and let ${}^{t}(K^{n'}_{comp})_0$ be the updated Gram matrix from time "0" (i.e., from the beginning till one last step) to time t. Similarly, the matrix ${}^{t+1}(y_{n'})_t$ indicates the update of the output matrix from time t to $t+1$ and ${}^{t}(y_{n'})_0$ indicates the update of output matrix till time t. The alignment factor at time $t+1$ and till time t can be given as ${}^{t+1}(A'(k^{n'}_{comp}, y_{n}, y^{T}_{n'}))_t$ and ${}^{t}(A'(k^{n'}_{comp}, y_{n'}y^{T}_{n'}))_0$, respectively.

$$ {}^{t+1}\left(A'(k^{n'}_{comp}, y_{n'}y^{T}_{n'})\right)_t < {}^{t}\left(A'(k^{n'}_{comp}, y_{n'}y^{T}_{n'})\right)_0 \tag{6.21} $$

This indicates that if the alignment factor deteriorates, then the parameter settings in Equation 6.18 should be set to a different value and the experiment should be run again. A significant variation in the alignment factor shows that the incoming nonstationary data is too large for correctly training it as either anomaly or normal. In this case, we have to divide and keep dividing the data into several small chunks of data until we find a suitable window size of nonstationary data that would be appropriate for training and yield the best results. This is discussed in the following subsection.

Now, because the training of small nonstationary datasets has to be sequential, let us consider the $d1$ data first. Compute the classes of the elements of the $d1$ data. Then, the input matrix at the current state, that is, at time $t+1$ should be ${}^{t+1}(P_r^{d1})_t = \left[{}^{t}(P_r^{n'})_0 \quad P_{d1}\right]$. Similarly, at time $t+1$, according to Equations 6.18–6.20, calculate ${}^{(t+1)}(B_{0r}^{d1})_t$, ${}^{(t+1)}(W_{0r}^{d1})_t$, ${}^{(t+1)}(M_{0r}^{d1})_t$, ${}^{(t+1)}(N_{0r}^{d1})_t$. The class separability of the composite data (data up to time t and $d1$ data) can be given as

$$ {}^{(t+1)}(J^{d1}(\alpha_r))_t = \frac{{}^{(t+1)}(\alpha_r^{d1})_t^T * {}^{(t+1)}(M_{0r}^{d1})_t * {}^{(t+1)}(\alpha_r^{d1})_t}{{}^{(t+1)}(\alpha_r^{d1})_t^T * {}^{(t+1)}(N_{0r}^{d1})_t * {}^{(t+1)}(\alpha_r^{d1})_t} \tag{6.22} $$

Therefore, ${}^{(t+1)}(\xi)_t = ({}^{(t+1)}(\lambda'_*)_t)/({}^{t}(\lambda_*)_0)$, where ${}^{(t+1)}(\lambda'_*)_t$ is the largest eigenvalue of the data (of the dominant kernel) received from time t to $t+1$, and ${}^{t}(\lambda'_*)_0$ is the largest eigenvalue of the dominant kernel calculated from time 0 to time t, that is, till the previous step. We test whether this subdataset $d1$ is either anomalous or normal. If ${}^{(t+1)}(\xi)_t$ is greater than a new threshold value η', then the incoming data is anomalous; otherwise, it is normal.

Minor Anomaly Case If the data $d1$ is anomalous, then we have to decide whether it is highly or less anomalous by making use of the residue factor (Eq. 6.19). If the residue factor ${}^{t+1}(rf^{d1})_t$ is less than 1, then the update of the $d1$ data is nonincremental. When the update is nonincremental, we have to find out and replace the trivial rows and columns in the dataset from the previous step and update them with the corresponding rows and columns from the new dataset $d1$. The combination coefficients in the previous step corresponding to the minimum of this difference are replaced by the corresponding vectors in the present step. This can be represented as follows:

Figure 6.4 shows the nonincremental update of combination vector at time $t+1$. Similarly, update ${}^{t+1}(P_r^{n'})_t$ and ${}^{t+1}(Q_r^{n'})_t$ and compute the new Gram matrix. Now, after determining the data-dependent kernels for $d1$ data, we have to determine the composite kernel using Equation 6.21. This case is summarized in the following steps:

Minor Anomaly Case

Step 1 Decide whether data $d1$ is highly or less anomalous using the Residue Factor. ${}^{(t+1)}(rf^{d1})_t = ({}^{(t+1)}(a_*^{d1})_t - {}^{t}(a_*)_0) \cdot \frac{{}^{(t+1)}(\lambda_*^{d1})_t}{{}^{t}(\lambda_*)_0}$

Step 2 Find out and replace the trivial rows and columns in the dataset from the previous step and update them with the corresponding rows and columns from the new dataset $d1$ by ${}^{t+1}(\Delta_{rk}^{d1})_t = {}^{t+1}(\alpha_r{}^{d1(l)})_t - {}^{t}(\alpha_{rk})_0$.

Step 3 Update ${}^{t+1}(P_r^{n'})_t$ and ${}^{t+1}(Q_r^{n'})_t$ to compute the new Gram matrix ${}^{(t+1)}(K_r^{n'})_t = {}^{t+1}(Q_r^{n'})_t * {}^{t+1}(P_r^{n'})_t * {}^{t+1}(Q_r^{n'})_t$.

Step 4 Determine the composite kernel as ${}^{t+1}(K_{\text{comp}}^{n'}(\rho))_t = \left[\sum_{i=1}^{p} \rho_i \left({}^{t+1}(K_r^{n'})_t\right)\right]_{({}^t n_0)\times({}^t n_0)}$.

Significant Anomaly Case If the data is a significant anomaly, we append the data so as to preserve the diverse information that may be used to classify the next incoming chunk of nonstationary data, that is, $d2$ as either anomaly or normal. The data-dependent kernel for highly anomaly $d2$ data can now give the matrices with the increased size as ${}^{(t+1)}(K_r^{n'})_t = {}^{t+1}(Q_r^{d2})_t * {}^{t+1}(P_r^{d2})_t * {}^{t+1}(Q_r^{d2})_t$. The remaining of the entire algorithm is summarized in the following steps.

Significant Anomaly Case

Step 1 Append the data to preserve the diverse information to classify the next incoming chunk of online data. Compute ${}^{t+1}(\alpha_r^{n'})_t = {}^{t+1}(\alpha_r^{d2})_t\ {}^{t+1}(Q_r^{n'})_t = \text{diag}({}^{t+1}(\alpha_r^{n'})_t) = {}^{t+1}(Q_r^{d2})_t,\ {}^{t+1}(P_r^{n'})_t = {}^{t+1}(P_r^{d2})_t$.

Step 2 Update the data-dependent kernel as: ${}^{(t+1)}(K_r^{n'})_t = {}^{t+1}(Q_r^{d2})_t * {}^{t+1}(P_r^{d2})_t * {}^{t+1}(Q_r^{d2})_t$

Step 3 Find a composite kernel that will yield optimum classification accuracy as: ${}^{t+1}(K_{\text{comp}}^{n'}(\rho))_t = \left[\sum_{i=1}^{p} \rho_i * {}^{t+1}(K_r^{n'})_t\right]_{(({}^t n_0)+l)\times(({}^t n_0)+l)}$.

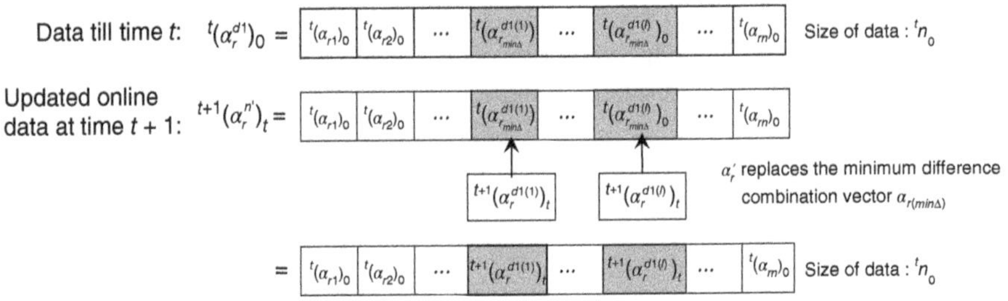

Figure 6.4 Nonincremental update of data at time $t + 1$. The data sequence with time index is evaluated for replacing the previous Gram matrix by the new time sequential datasets.

6.5.2 Anomaly Prediction of Long-Time Sequential Trajectories

k-Step prediction uses the existing data sequence to predict the next k-step value using ARMA model [8], in which only the output estimates are used, and the uncertainty induced by each successive prediction is not accounted for. In order to avoid this limitation, we propose the framework of "predictive LACK" by extending the following composite kernels for the k-step in ahead of time index, that is, $t + k$, instead of one step of $t + 1$:

$$^{t+k}\left(K^{n'}_{\text{comp}}(\rho)\right)_t = \left[\sum_{i=1}^{p} \rho_i * {}^{t+k}(K^{n'}_r)_t\right] \tag{6.24}$$

Equation 6.24 corresponds to the stage of cancer that is used for the prediction of normality and abnormality via the extension of LACK described in Section 6.4. The long data sequences are considered by changing different size of horizontal time window.

The anomaly prediction in a clinical application is called cancer staging, which is given by a number ranging from 0 to IV, with IV having more progression, as shown in Fig. 6.5. The cancer stage often takes into account the size of a tumor; however, several other factors are also concerned with the stage of the cancer, such as whether it has invaded adjacent organs, how many lymph nodes it has metastasized to, and whether it has spread to distant organs. Staging of cancer is the most important predictor of survival, and cancer treatment is primarily determined by staging.

The stage of a cancer is used for the quantitative values for prediction using LACK. Rather than simple and traditional binary classification of TP/FP of tumor images, the anomaly degree is evaluated by the kernel Gram matrix. The cancer staging is then extended into the prediction framework that describes whether it has spread to distant organs. LACK adapts the staging of cancer, the most important predictor of survival, so that cancer treatment planning can be determined by the predicted stage of cancer.

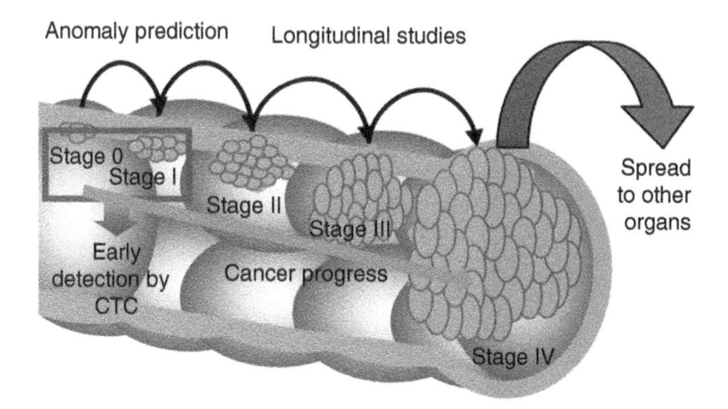

Figure 6.5 A conceptual illustration of colorectal cancer staging. The degree of cancer progress is indexed by stages 0–IV, corresponding to the size of the tumor.

The proposed framework, called predictive LACK, consists of the following five steps:

Predictive LACK

Step 1 Regroup 3010 datasets for applying stationary composite kernel matrix algorithm: $k_r(x_i, x_j) = q_r(x_i)q_r(x_j)p_r(x_i, x_j)$, $K_{\text{comp}}(\rho) = \sum_{t=1}^{p} \rho_i Q_i P_i Q_i$, and $J = \left(\text{tr}\left(\sum_r S_{br}\right)\right) / \left(\text{tr}\left(\sum_r S_{wr}\right)\right)$.

Step 2 Apply alignment factors to determine whether divide datasets into small subsets: ${}^{t+1}(A\prime(k^{n'}_{\text{comp}}, y_{n'}y^T_{n'}))_t < {}^{t}(A\prime(k^{n'}_{comp}, y_{n'}y^T_{n'}))_0$, ${}^{(t+1)}(J^{d1}(\alpha_r))_t = \frac{{}^{(t+1)}(\alpha_r^{d1})_t^T * {}^{(t+1)}(M_{0r}^{d1})_t * {}^{(t+1)}(\alpha_r^{d1})_t}{{}^{(t+1)}(\alpha_r^{d1})_t^T * {}^{(t+1)}(N_{0r}^{d1})_t * {}^{(t+1)}(\alpha_r^{d1})_t}$.

Step 3 Calculate LACK for sequences of time horizontal window, starting from 1 to k: ${}^{(t+1)}(K_r^{n'})_t = {}^{t}(Q_r^{n'})_0 * {}^{t}(P_r^{n'})_0 * {}^{t}(Q_r^{n'})_0$.

Step 4 Compute kernel matrix reconstruction accuracy using $\text{MErr} = (1/n)\sum_{i=l+1}^{n} \lambda_i$, with composite kernel matrix to convert the k-composited matrix value to the synthesized measurement value.

Step 5 Find out the cancer stage corresponding to the synthesized measurement data.

This entire process is summarized in the algorithm flow shown in Fig. 6.6.

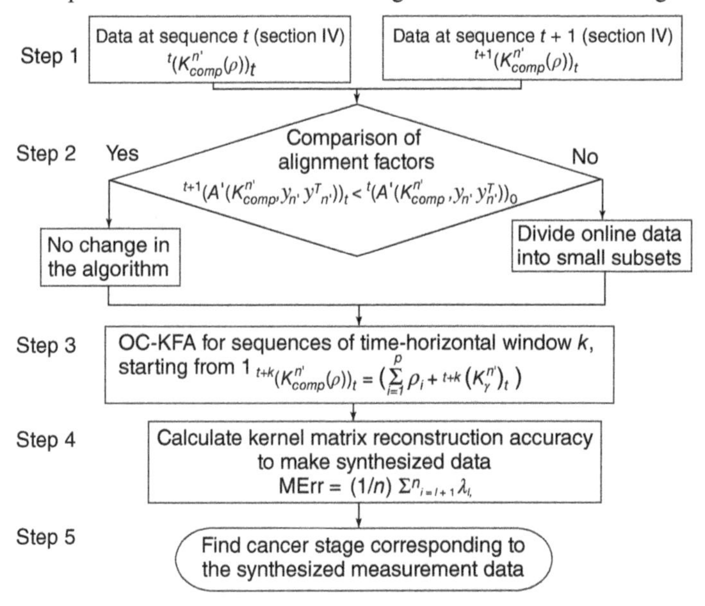

Figure 6.6 The training of nonstationary datasets is acquired over long-time sequences to divide the huge dataset by alignment factors, and to compute the cancer stage corresponding to synthesized measurement data.

6.6 CLASSIFICATION RESULTS

The proposed Predictive LACK is evaluated using CT image datasets of colonic polyps composed of true positive (TP) and false positive (FP) polyps detected by our CAD system [4]. The following subsections will evaluate cancer datasets (Section 6.6.1), selection of optimum kernel and composite kernel for stationary data (Section 6.6.2), comparisons with other kernel learning methods (Section 6.6.3), and anomaly detection for the nonstationary data (Section 6.6.1).

6.6.1 *Cancer Datasets*

For the stationary experiment, we obtained studies of 146 patients who had undergone a colon-cleansing regimen that is the same as that of optical colonoscopy. Each patient was scanned in both supine and prone positions, resulting in a total of 292 CT studies. The volumes of interest (VOIs) representing each polyp candidate have been calculated as follows: The CAD scheme provided a segmented region for each candidate. The center of a VOI was placed at the center of mass of the region. The size of the VOI was chosen so that the entire region was covered. The resampling was carried out using VOIs with dimensions $12 \times 12 \times 12$ voxels to build Dataset1 that consists of 29 true polyps and 101 false polyps. For the rest of the datasets, the VOI was re-sampled to $16 \times 16 \times 16$ voxels. The VOIs computed were Dataset1 (29 TPs and 101 FPs), Dataset2 (54 TPs and 660 FPs), Dataset3 (16 TPs and 925 FPs), and Dataset4 (11 TPs and 2250 FPs).

For the nonstationary experiment, we used a large dataset that has a total of 3749 CT datasets. For acquisition of these datasets, helical single-slice or multislice CT scanners were used, with collimation of 0.5 mm, reconstruction intervals of 0.5 mm, X-ray tube currents of 50–260 mA, and voltages of 120–140 kVp. The in-plane pixel size was 0.5 mm, and the CT image matrix size was 512×512. As shown in Table 6.1, we have divided the datasets into four groups. There were 368 normal cases and 54 abnormal cases with colonoscopy-confirmed polyps larger than 48 mm.

Table 6.1 shows the arrangement of stationary training and testing sets, DataSet1, DataSet2, DataSet3, and DataSet4, as well as the nonstationary training and anomaly

Table 6.1 Arrangement of datasets

	Number of vectors in training set			Number of vectors in testing set		
Datasets	TP	FP	Total	TP	FP	Total
Stationary Set1	21	69	90	8	32	40
Stationary Set2	38	360	398	16	300	316
Stationary Set3	10	500	510	6	425	431
Stationary Set4	7	1050	1057	4	1200	1204
Nonstationary Set1	15	403	418	19	500	519
Nonstationary Set2	20	503	423	28	600	628
Nonstationary Set3	25	706	731	29	900	929

testing sets, Nonstationary Set1, Nonstationary Set2, Nonstationary Set3. Instead of using the cross validation, we randomly divided the entire datasets into training and testing.

6.6.2 Selection of Optimum Kernel and Composite Kernel for Stationary Data

We used the method proposed in Section 6.3.1 to create four different data-dependent kernels, selected the kernel that best fit the data, and gave optimum classification accuracy for the stationary data. We determined the optimum kernel depending on the eigenvalue that yielded maximum separability. Table 6.2 indicates the eigenvalues λ and parameters of four kernels for each dataset calculated in Equation 6.14.

The kernel with the maximum value is highlighted for each stationary dataset in Table 6.2. According to the order of eigenvalue, we selected the two largest kernels to form the composite kernel. Taking Dataset1 as an example, we combined RBF and Laplace to form the composite kernel. We observed that each database has different combinations for the composite kernels. The composite coefficients of the two most dominant kernels in Table 6.2 are listed in Table 6.3.

Table 6.3 shows how the two kernel functions are combined according to the composite coefficients listed. These composite coefficients were obtained using Equation 6.17 in Section 6.5.2. For all of the datasets, the most dominant kernel was

Table 6.2 Eigenvalues of four kernels for stationary datasets

Datasets	Parameters and λ for linear kernel (Eq. 6.1)	Parameters and λ for polynomial kernel (Eq. 6.2)	Parameters and λ for RBF kernel (Eq. 6.3)	Parameters and λ for Laplace kernel (Eq. 6.4)
Stationary Set1	$\lambda = 10.66$	$\lambda = 10.25$	$\lambda = \mathbf{14.13}$	$\lambda = \mathbf{12.41}$
		$d = 1.2$, Offset = 2	$\sigma = 4.12$	$\sigma = 0.9$
Stationary Set2	$\lambda = 102.08$	$\lambda = \mathbf{105.91}$	$\lambda = \mathbf{116.64}$	$\lambda = 80.57$
		$d = 1$, Offset = 4	$\sigma = 5.29$	$\sigma = 3.5$
Stationary Set3	$\lambda = \mathbf{57.65}$	$\lambda = 51.35$	$\lambda = \mathbf{74.55}$	$\lambda = 30.23$
		$d = 1.4$, Offset = 1	$\sigma = 5.65$	$\sigma = 1.0$
Stationary Set4	$\lambda = 72.41$	$\lambda = \mathbf{83.53}$	$\lambda = \mathbf{124.13}$	$\lambda = 56.35$
		$d = 0.8$, Offset = 2	$\sigma = 4$	$\sigma = 2.5$

Table 6.3 The value of $\hat{\rho}$ for each of the composite kernels

Datasets	Two most dominant kernels	Linear combination of kernels
Stationary Set1	RBF and Laplace	$\rho_1 = 0.98, \ \rho_2 = 0.14$
Stationary Set2	RBF and polynomial	$\rho_1 = 0.72, \ \rho_2 = 0.25$
Stationary Set3	RBF and linear	$\rho_1 = 0.98, \ \rho_2 = 0.23$
Stationary Set4	RBF and polynomial	$\rho_1 = 0.91, \ \rho_2 = 0.18$

Table 6.4 Classification accuracy and mean square error of stationary data

Datasets	Mean square reconstruction error of stationary data with composite kernel, %	Classification accuracy of stationary data with composite kernel, %
Stationary Set1	1.0	90
Stationary Set2	9.64	94.62
Stationary Set3	6.25	98.61
Stationary Set4	14.03	99.67

the RBF kernel, where as the second most dominant kernel kept varying. As a result, the contribution of the RBF kernel was higher when compared to other kernels in forming a composite kernel.

The classification accuracy and mean square reconstruction error for the stationary datasets are shown in Table 6.4. The mean square reconstruction error in the case of stationary data using AKFA algorithm was calculated by $\text{Err}_i = ||\Phi_i - \Phi'_i||^2$. The classification accuracy was calculated by (TP+TN)/(TP+TN+FN+FP), where TN, TP, FP, and FN stands for true negative, true positive, false positive, and false negative, respectively.

Table 6.4 shows that Stationary Set4 has the highest classification accuracy even if the mean square reconstruction has the largest error. This demonstrated that composite kernel approach can well handle this type of unbalanced datasets.

Figure 6.7 also statistically shows that we obtained very good classification accuracy and ROC performance by using a composite data-dependent kernel over all of the stationary datasets.

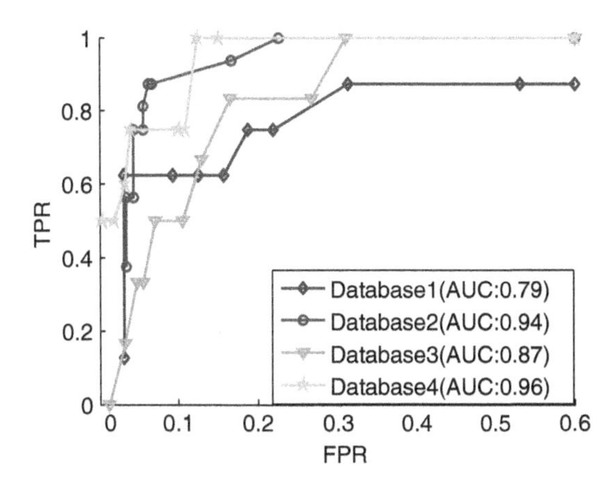

Figure 6.7 The ROC performance of stationary databases 1–4 using composite KFA with k-NN. AUC of four databases shows very competitive results for the other kernel learning methods shown in Fig. 6.8.

Table 6.5 Overview of framework methodology

Methods	Kernel formation	Advantages
LACK	Combine the most dominant kernels determined by maximizing the alignment measure for detection and adaptation to changes in underlying data distributions	Used for the prediction of anomaly status and cancer stage by iteratively constructing a high-dimensional feature space, while also maximizing a variance condition for the nonlinearly transformed samples
MKL [13]	Identify the key properties of the existing MKL algorithms in order to construct a taxonomy, highlighting similarities and differences between them	Combination into one step with a criterion, and handling a larger number of kernels with different parameter settings for various datasets

6.6.3 Comparisons with Other Kernel Learning Methods

LACK and the existing online multiple kernel learning (MKL) studies [13] have been compared in Table 6.5 as follows.

In the performance comparison, we compared other popular machine-learning techniques, called support vector machine (SVM) [7] shown in Table 6.6. The selection of kernel function is identical to that of the KPCA cases of the chosen single (Table 6.2) and composite (Table 6.3) kernels.

Table 6.6 shows, compared to Table 6.4 using composite kernel, that SVM had a better classification accuracy for two datasets (Colon1, Colon2) than KPCA, or sparse kernel feature analysis (SKFA), or AKFA with single and composite kernels, but the other two datasets (Colon3, Colon4) achieved lower classification accuracy.

Table 6.6 Classification Accuracy to compare other multiple kernel methods

Datasets	Proposed composite KFA for SVM	Parameters for two selected kernels	Single KFA for SVM	Parameters for gaussian RBF
Stationary Set1	100.0	Slack = 0.006 $\sigma_1 = 0.37, \sigma_2 = 0.37$ $D = 0.55$, Offset = 0.825	100.0	Slack = 0.25 $\sigma = 0.5$
Stationary Set2	94.94	Slack = 0.007 $\sigma_1 = 0.73, \sigma_2 = 0.73$ $D = 0.46$, Offset = 0.69	94.30	Slack = 0.004 $\sigma = 0.82$
Stationary Set3	86.49	Slack = 0.000002 $\sigma_1 = 0.28, \sigma_2 = 0.28$ $D = 0.82$, Offset = 1.23	83.78	Slack = 0.000002 $\sigma = 0.64$
Stationary Set4	96.77	Slack = 0.0001 $\sigma_1 = 0.1, \sigma_2 = 0.1$ $D = 1$, Offset = 1	96.34	Slack = 0.005 $\sigma = 0.01$

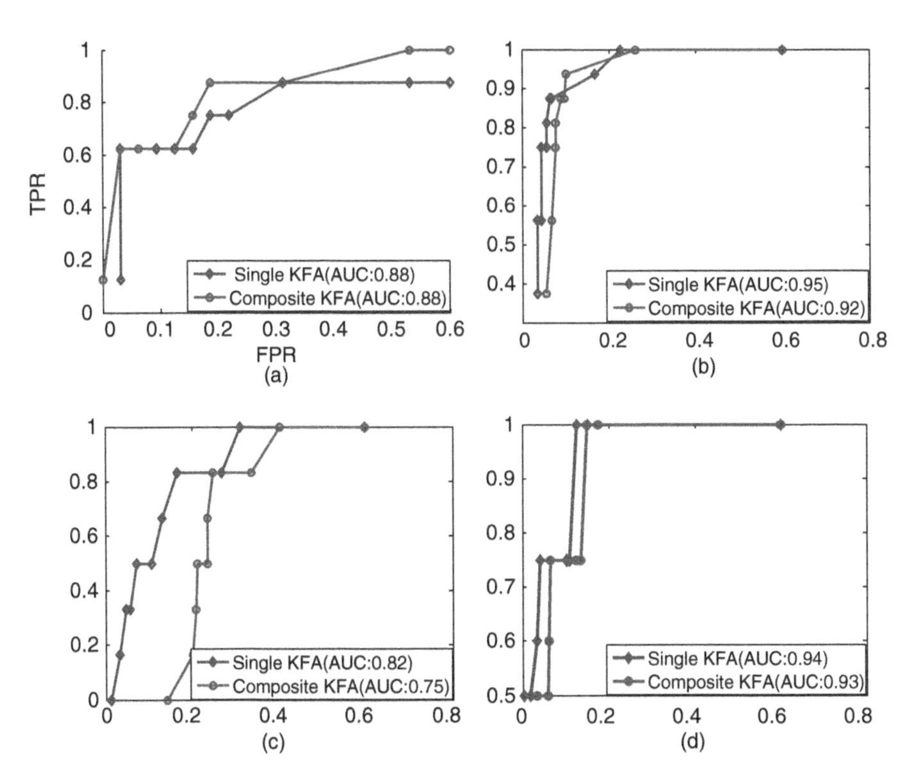

Figure 6.8 The ROC performance of single KFA versus composite KFA for SVM. Composite KFA approaches work on the top of SVM, showing good results of four databases ((a) Database1, (b) Database2, (3) Database3, and (d) Database4) with high AUC vales.

In Fig. 6.8, we can see that the proposed composite kernel feature analysis (KFA) for SVM of area under curve (AUC) in Database2, Database3, and Database4 outperformed the performance of single KFA for SVM. This result indicates the proposed principal composite kernel feature analysis was very competitive to the well-known classifier for different datasets.

PROTEIN [44] composed of 10 different feature representations and two kernels for 694 instances (311 for training and 383 for testing). PENDIGITS [45] composed of four different feature representations for 10,992 instances (7494 for training and 3498 for testing). Two binary classification problems are generated from the PENDIGITS data by even digits and odd digits. ADVERT [46] composed of five different feature representations (3279 images). The classification task is to predict whether an image is an advertisement. These three datasets were used to compare SimpleMKL, ABMKSVM, RBMKSVM, and LMKL in the experiment [13]. SimpleMKL used linear and polynomial ($p = 2$) as the basic kernel. ABMKSVM (convex) solves the QP problem ($p = 2$), RBMKSVM (product) trains an SVM with the product of the combined kernels ($p = 2$), and LMKL (sigmoid) uses the

Table 6.7 Classification accuracy to compare other multiple kernel methods using UCI machine learning repository [47]

Datasets	Simple MKL [48]	ABMKSVM [49]	RBMKSVM [50]	LMKL [51]	LACK
Stationary Set1	99.13	100.0	99.13	99.13	100.0
Stationary Set2	*95.13*	93.13	94.94	91.35	94.94
Stationary Set3	*84.91*	85.32	80.07	86.49	86.49
Stationary Set4	95.56	94.81	93.36	95.56	96.77
PROTEIN [44]	76.34	77.99	75.18	81.07	82.06
PENDIGITS [45]	93.29	93.42	95.90	97.47	96.31
ADVERT [46]	96.37	95.67	93.25	95.58	96.02

sigmoid gating model ($p = 2$). In Table 6.7, the performance of the proposed LACK confirmed its superiority.

6.6.4 Anomaly Detection for the Nonstationary Data

We evaluated the proposed LACK to tune the selection of appropriate kernels when new nonstationary data becomes available. We divided the dataset equally into 10 sets to form the nonstationary data stream in Table 6.7, which shows the size of different nonstationary batches of data for each colon cancer dataset. Note that 10 nonstationary data sequences were generated from 3010 larger sequential datasets with 600 anomaly datasets and 2410 normal datasets. Each nonstationary dataset was randomly divided into subsets. For example, the first sampled nonstationary dataset had 960 datasets, which was used for 10 small chunks (each chunk with 20 anomaly vs 76 normal). Others had another 1000, and 1050, in the total of three nonstationary datasets. After we tentatively form the input matrices for four different kernels, we used Equations 6.16 and 6.17 to find the dominant kernels for the new data and the previous stationary data. These results are summarized in Table 6.8.

The kernel with the maximum eigenvalue is highlighted for each nonstationary dataset sequence in Table 6.8; we can see that the RBF kernel was always the most dominant kernel in all dataset, and the second dominant kernel kept varying.

Table 6.9 shows the results of detection of anomaly status using the proposed criterion of class separability $\xi = \lambda_*' / \lambda_*$ (Eq. 6.18) and residue factor "*rf*" (Eq. 6.19). In Section 6.5, there were three statuses of the anomaly identification: normal, minor anomaly, and significant anomaly, corresponding to the anomaly degree. Using the proposed LACK methods, the classification accuracy and AUC were calculated for each dataset for nonstationary sequences. The classification performance was similar to Stationary datasets shown in Table 6.4. Each sequence for LACK was cascaded into one long sequence for the next experiment of large nonstationary sequential dataset. The anomaly detection performance was relatively high; thus these algorithms have the potential to be used in a preclinical setting of patients' diagnosis.

Table 6.8 Eigenvalues λ of four different kernels for nonstationary data sequences #1–#10 for the base nonstationary Sets #1–#3

Datasets	Kernel	#1	#2	#3	#4	#5	#6	#7	#8	#9	#10
Nonstationary Set1	Linear	18.9	8.43	25.7	**11.8**	**31.6**	14.9	**27.8**	**19.0**	**29.3**	**16.8**
	Poly	18.3	8.41	24.8	9.59	28.5	13.2	24.1	14.4	28.5	16.2
	RBF	**20.5**	**8.76**	**28.7**	9.43	30.1	**16.3**	25.5	14.1	27.9	13.3
	Laplace	11.7	7.40	24.5	11.4	27.9	14.0	23.2	17.8	21.8	14.8
Nonstationary Set2	Linear	3.23	6.26	3.91	7.24	9.78	9.13	4.10	7.02	9.71	7.43
	Poly	3.17	6.08	4.79	7.61	9.17	8.09	4.79	7.81	9.78	8.81
	RBF	**3.54**	**6.41**	**4.97**	**7.92**	**10.2**	**9.21**	**6.28**	**8.06**	**10.9**	**8.89**
	Laplace	2.80	5.34	3.58	6.23	8.97	7.26	3.85	6.37	8.29	6.69
Nonstationary Set3	Linear	21.3	34.8	18.8	22.1	25.0	35.9	24.1	24.1	24.1	27.8
	Poly	33.7	43.6	19.3	22.8	26.1	37.1	25.8	27.9	28.6	29.6
	RBF	**34.1**	**45.7**	**20.3**	**24.7**	**28.9**	**38.6**	**28.0**	**28.7**	**29.1**	**30.6**
	Laplace	21.8	20.9	17.4	20.8	18.3	29.7	21.8	21.5	16.8	23.1

The bold fonts indicate the highest values.

Table 6.9 Anomaly detection with classification accuracy and AUC for nonstationary data sequences #1–#10, and for the base nonstationary sets #1–#3

Datasets	Item	#1	#2	#3	#4	#5	#6	#7	#8	#9	#10
Nonstationary Set1	Update	N	N	M	N	N	S	M	N	S	M
	Accuracy	93.7	93.7	94.1	94.1	94.1	92.5	94.6	94.6	98.7	91.4
	AUC	87.2	87.2	89.3	89.3	89.3	88.6	89.1	89.1	90.1	85.7
Nonstationary Set2	Update	N	N	N	N	M	S	S	M	M	N
	Accuracy	97.2	97.2	97.2	97.2	98.1	97.9	96.9	98.0	92.5	92.5
	AUC	94.2	94.2	94.2	94.2	94.1	93.8	95.0	93.7	88.6	88.6
Nonstationary Set3	Update	N	M	N	S	N	N	M	N	S	M
	Accuracy	94.6	95.3	95.3	93.8	93.8	93.8	92.6	92.6	98.5	91.6
	AUC	89.6	91.2	91.2	87.9	87.9	87.9	91.8	91.8	94.1	87.4

N indicates normal, M indicates minor anomaly, S indicates significant anomaly.

6.7 LONGITUDINAL PREDICTION RESULTS

The following subsections will evaluate the anomaly prediction performance of longitudinal sequential trajectories described in Section 6.7. We will analyze this in Sections 6.7.1–6.7.3.

6.7.1 Large Nonstationary Sequential dataset for Anomaly Detection

We applied the LACK method mentioned in Section 6.5.1 by extending the anomaly detection of small limited cases to the larger nonstationary sequential datasets using

Equation 6.22 for the next step to detect anomaly cases. Training and testing for classification of labeled anomalies (anomaly or normal) handled both anomaly/normal data with each stage indexing. These anomaly cases are shown in Fig. 6.9, and these normal cases are shown in Fig. 6.10.

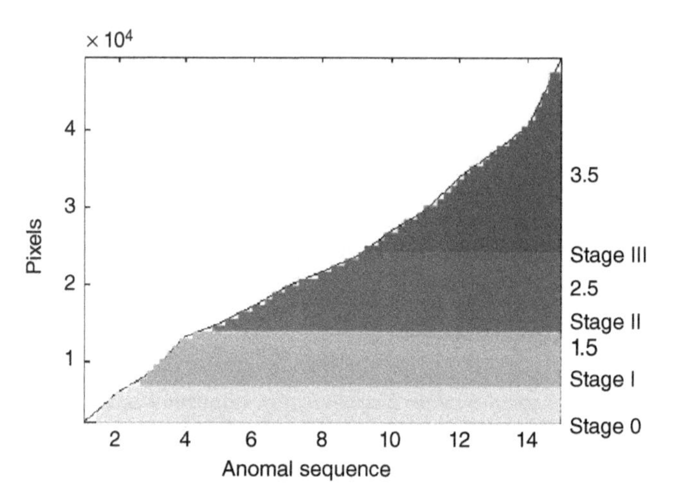

Figure 6.9 Labeling "anomaly" with cancer stage for all TP cases of time indexing corresponding to the segmented polyp pixels. Area of pixel histogram (APH) is used for assigning each cancer stage from 0 to III.

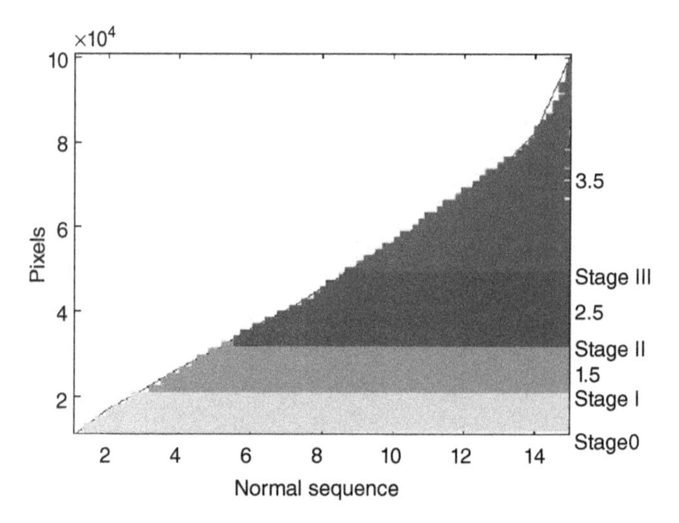

Figure 6.10 Labeling "normal" with cancer stage for all FP cases of time indexing corresponding to the segmented polyp pixels. Area of pixel histogram (APH) represent each cancer stage from 0 to III.

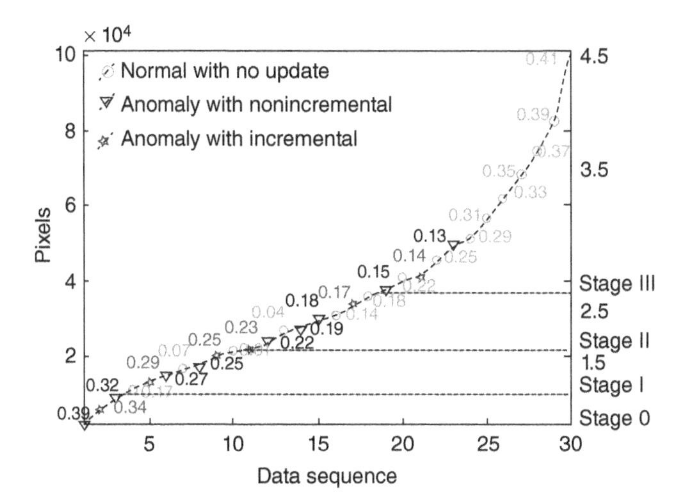

Figure 6.11 Transient from normal to anomaly among longitudinal sequential datasets. The indexing of cancer stages is based on Figs. 6.9 and 6.10. The 15 cases of transition from normal to anomaly are shown.

Figures 6.9 and 6.10 make the conversion from pixel histogram to the cancer stage. We then applied anomaly detection to the entire nonstationary sequential datasets using transition of Fig. 6.11.

In Fig. 6.11, we analyzed the longitudinal sequential datasets transient from normal to anomaly among cancer stage index 0–III. All transitions in Fig. 6.11 included 15 normal and anomaly cases, so we trained all the transitional cases among the entire combinatory cases of ${}_{30}C_2 = 30 * 29/2 = 435$. All 30 nonstationary data sequences that indexed longitudinal cases of these normal-anomaly transitions were used for training, and the current cancer indexing (normal) would be the next anomaly case by introducing the post-test probability [9]. After all the normal-anomaly transitions were trained, then a couple of new nonstationary data sequences with a certain time index were used for determining, in the future data sequence index, how much post-test probabilities the anomaly case would be expected to have, based on the next several nonstationary sequences.

Figure 6.12 shows that post-test probability, which corresponds to the transitional nonstationary data sequence index. The probabilities were calculated to meet the following pre-test odds by the likelihood ratio. The post-test odds represented the chances that your patient had a disease. It incorporates the disease prevalence, the patient pool, specific patient risk factors (pre-test odds), and information about the diagnostic test itself (the likelihood ratio) [11].

Figure 6.12 shows cancer stage 3.2 had the lowest post-test probability, indicating the highest concern of further diagnosis. Normal cases with four low post-test probability of the cancer stage 1.1, 1.7, 1.9, and 2.2, had certain higher risk factors. Even a larger cancer stage (corresponding to large pixels) had many normal cases with good post-test probability numbers, mainly due to the FP cases.

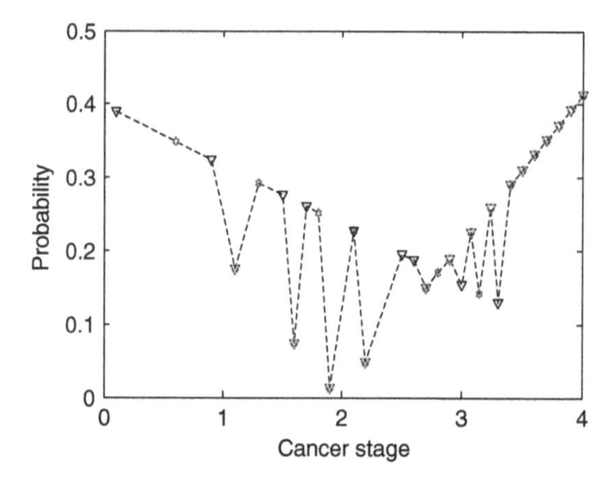

Figure 6.12 Post-probabilities of anomaly cases among 15 longitudinal sequential datasets. Based on Fig. 6.11 this plot reconfigured the axis to see the sorting order how cancer stage calculated post-probabilities of anomaly cases.

6.7.2 Time Horizontal Prediction for Risk Factor Analysis of Anomaly Long-Time Sequential Trajectories

We evaluated the predictive performance on the basis of the anomaly trajectories, which means the time horizontal transition of earlier low cancer stage (mostly called normal case) to the later larger cancer stage, using all five long-time nonstationary sequential datasets in Table 6.1. Note that, in this subsection, a large nonstationary dataset is much longer than the nonstationary datasets previously described.

Figure 6.13 shows the variable of time horizontal window, starting from each frame up to 3613 frames to analyze the risk of cancer stages. The representative size of the horizontal time window was set to predict cancer stages in advance. For example, in Fig. 6.13, these predicted window sizes k were 1, 5, 10, 50, and 100 time index for the predicted values. The k-step prediction method of Fig. 6.13(a) shows larger error than LACK prediction method of Fig. 6.13(b), especially when the predicted window size was 50 and 100.

These prediction results of Fig. 6.13 are summarized in Table 6.10. The algorithms of k-step and the proposed LACK prediction run into all 3613 data sequences, which included a finite number of sequences called low cancer stage (I–II) and high cancer stage (II–III). For the comparison of prediction performance, a normalized metric was used, that is, the normalized root mean squared error (NRMSE) between the predicted cancer stage and the defined cancer stage used in Figs. 6.9 and 6.10.

Figure 6.13 and Table 6.10 show that NRMSE of II–III was larger than NRMSE of I–II. The traditional k-step and proposed LACK prediction methods were both efficient in handling nonstationary data over small window-size sequences with modest NRMSE. The prediction performance of the long prediction time window size (50, 100), indicated that the prediction of the subsequent larger cancer stages of data using LACK was advantageous over a longer prediction horizontal window size.

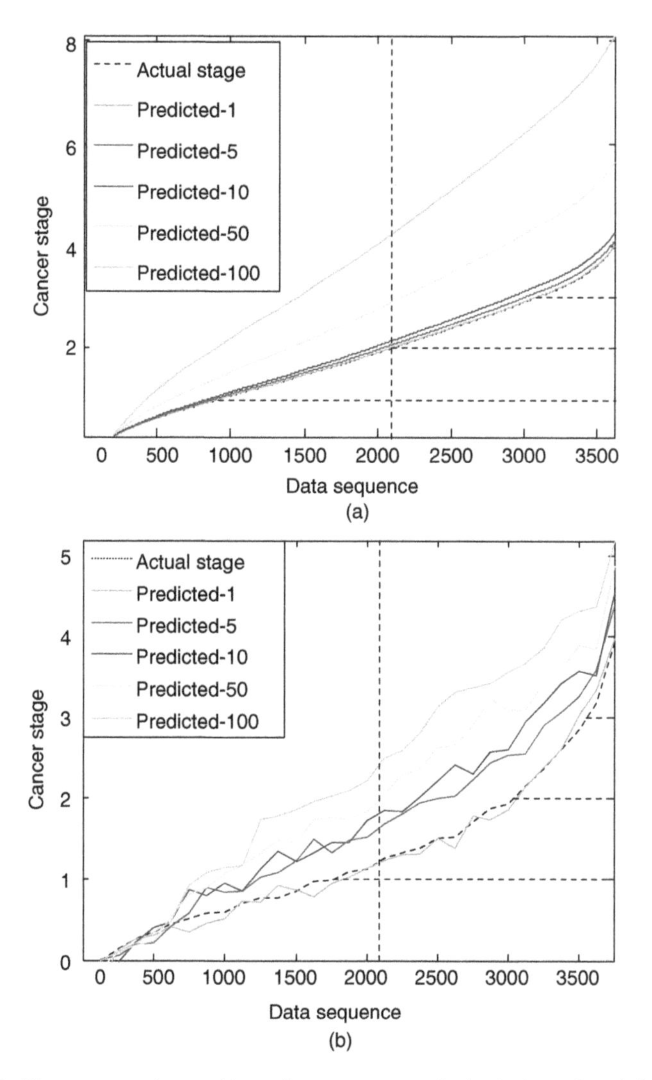

Figure 6.13 The cancer stage of long-time sequence trajectories using (a) k-step prediction, (b) predictive LACK. When the horizontal time window increased 1, 5, 10, 50, and 100 ahead of the time of data sequence, the predicted values were off from the actual cancer stage value. (a) k-Step and (b) predictive LACK performed similar with small predicted window sizes k such as $k = 1 - 10$. If $k = 50$ or 100 (under the larger prediction time frame), LACK performed much better than K-step.

6.7.3 Computational Time for Complexity Evaluation

Table 6.11 shows the computational cost of the proposed LACK over all four stationary and three nonstationary data without discarding it. In Table 6.11, the following four representative modules on LACK are listed: "Kernel selection" indicates Section 6.3 for the results of Tables 6.2 and 6.3. "Cancer classification" indicates the

Table 6.10 Average error of prediction using several representative prediction window sizes

Horizontal window time size	Low cancer stages (I–II)	High cancer stages (II–III)	Entire cancer stages
k-step 1	0.0029	0.0060	0.0067
k-step 5	0.0294	0.0603	0.0671
k-step10	0.0660	0.1356	0.1508
k-step 50	0.4125	0.8534	0.9478
k-step 100	1.0157	2.1205	2.3509
LACK step 1	0.1041	0.0904	0.0980
LACK step 5	0.2777	0.4935	0.3934
LACK step 10	0.3348	0.6860	0.5286
LACK step 50	0.5004	1.0387	0.7981
LACK step 100	0.7169	1.4919	1.1458

Table 6.11 Computational time (millisecond) of each module for each dataset

Datasets	Kernel selection	Cancer classification	Normal detection	Anomaly prediction
Stationary Set1	5.7	0.81	NA	NA
Stationary Set2	31.9	4.1	NA	NA
Stationary Set3	43.2	4.9	NA	NA
Stationary Set4	63.4	6.1	NA	NA
Nonstationary Set1	19.8	2.0	0.62	0.61
Nonstationary Set2	22.3	2.1	0.71	0.72
Nonstationary Set3	30.1	2.4	1.09	1.08
Average	30.9	3.2	0.80	0.80

results of Table 6.4 for ROC Figs. 6.7–6.9. "Anomaly detection" indicates Section 6.4 for the results of Figs. 6.10–6.13. "Anomaly prediction" indicates Section 6.5 for the results of Fig. 6.13. We used a PC with an Intel i7 3.16 GHz CPU and 16 GB of RAM.

Table 6.11 shows that, for the average of all seven different types of datasets, the module of Kernel Selection along with Stationary sets 1–3 resulted in the highest computational load with an average of 30.9 ms among the four modules. The time reduction of nonstationary learning was 33.3% less than stationary learning. Thus, our proposed LACK computationally was very efficient for nonstationary leaning. When adding anomaly detection and prediction, the overall computing cost did not change much due to the light modules.

6.8 CONCLUSIONS

This chapter proposed a novel method of detection and prediction from the long-time anomaly trajectories of nonstationary data sets using stationary training data. This method, called longitudinal analysis of composite kernels (LACK), was a faster and more efficient feature extraction algorithm, derived from the KPCA, for the detection of polyps on CT colonographic images. The polyp classification experiment showed that LACK offered equivalent classification performance of nonstationary data compared to that of stationary data. The k-step extension of LACK has the potential to predict the cancer stage index and yields high detection performance regarding the status of anomalies. Such an effective predictive scheme has the potential to make CT a viable option for screening large patient populations, resulting in early detection of colon cancer, and leading to a reduction in colon-cancer-related deaths through cancer stage analysis.

REFERENCES

1. J. J. Wood, R. J. Longman, and N. Rooney. "Colonic vascular anomalies and colon cancer in neurofibromatosis: Report of a case," *Dis. Colon Rectum*, vol. 51, no. 3, pp. 360–362, 2008.
2. P. Lefere, S. Gryspeerdt, and A.L. Baert, *Virtual colonoscopy: A practical guide*. Springer, New York, NY, 2009.
3. H. Yoshida, Y. Masutani, P. MacEneaney, D. T. Rubin, and A. H. Dachman, "Computerized detection of colonic polyps at CT colonography on the basis of volumetric features: Pilot study," *Radiology*, vol. 222, pp. 327–336, 2002.
4. H. Yoshida and J. Näppi, "Three-dimensional computer-aided diagnosis scheme for detection of colonic polyps," *IEEE Trans. Med. Imaging*, vol. 20, pp. 1261–1274, 2001.
5. J. Näppi and H. Yoshida, "Fully automated three-dimensional detection of polyps in fecal-tagging CT colonography," *Acad. Radiol.*, vol. 14, pp. 287–300, 2007.
6. H. Yoshida and J. Näppi, "CAD in CT colonography without and with oral contrast agents: Progress and challenges," *Comput. Med. Imaging Graph.*, vol. 31, pp. 267–284, 2007.
7. W.S. Zheng, J.H. Lai, and P.C. Yuen, "Penalized pre-image learning in kernel principal component analysis," *IEEE Trans. Neural Netw.*, vol. 21, no. 4, pp. 551–570, 2010.
8. Y. Zhaoa, J. Chua, H. Sua, and B. Huang, "Multi-step prediction error approach for controller performance monitoring," *Control Eng. Pract.*, vol. 18, no. 1, pp. 1–12, 2010.
9. H. C. van Woerden, A. Howard, M. Lyons, D. Westmoreland, and M. Thomas, "Use of pre- and post-test probability of disease to interpret meningococcal PCR results," *Eur. J Clin. Microbiol. Infect. Dis.*, vol. 23, no. 8, 658–660, 2004.
10. Y. Tao, R. Cheng, X. Xiao, W. K. Ngai, B. Kao, and S. Prabhakar, "Indexing multi-dimensional uncertain data with arbitrary probability density functions," in *Proceedings of 31st International Conference on Very large Data Bases*, Trondheim, Norway, 30 August–02 September, 2005.
11. Chen, H., Chen, and J., Kalbfeisch, J.D. "A modified likelihood ratio test for homogeneity in the finite mixture models," *J. Roy. Stat. Soc. B.*, vol. 63, pp. 19–29, 2001.
12. H. Xiong, M.N.S. Swamy, and M.O. Ahmad, "Optimizing the data dependent kernel in the empirical feature space," *IEEE Trans. Neural Netw.*, vol. 16, pp. 460–474, 2005.
13. M. Gönen and E. Alpaydin, "Multiple kernel learning algorithms," *J. Mach. Learn. Res.*, vol. 12, pp. 2211–2268, 2011.
14. B. J. Kim, I. K. Kim, and K. B. Kim, "Feature extraction and classification system for nonlinear and online data," in *Advances in Knowledge Discovery and Data mining, Proceedings*, Sydney, Australia, May 26–28, vol. 3056, pp. 171–180, 2004.

15. W. Zheng, C. Zou, and L. Zhao, "An improved algorithm for kernel principal component analysis," *Neural Process. Lett.*, vol. 22, pp. 49–56, 2005.
16. J. Kivinen, A. J. Smola, and R. C. Williamson, "Online learning with kernels," *IEEE Trans. Signal Process.*, vol. 52, no. 8, pp. 2165–2176, 2004.
17. S. Ozawa, S. Pang, and N. Kasabov, "Incremental learning of chunk data for online pattern classification systems," *IEEE Trans. Neural Netw.*, vol. 19, no.6, pp. 1061–1074, 2008.
18. K.L. Chan, P. Xue, L. Zhou, "A kernel-induced space selection approach to model selection in KLDA," *IEEE Trans. Neural Netw.*, vol. 19, no. 12, pp. 2116–2131, 2008.
19. Y. M. Li, "On incremental and robust subspace learning," *Pattern Recogn.*, vol. 37, no. 7, pp. 1509–1518, 2004.
20. Y. Kim, "Incremental principal component analysis for image processing," *Opt. Lett.*, vol. 32, no. 1, pp. 32–34, 2007.
21. V. N. Vapnik, *The nature of statistical learning theory*, 2nd ed. New York: Springer, 2000.
22. R. O. Duda, P. E. Hart, and D. G. Stork. *Pattern classification*, 2nd ed. Hoboken, NJ: John Wiley & Sons Inc., 2001.
23. Y. Tan, J. Wang, "A support vector machine with a hybrid kernel and minimal Vapnik-Chervonenkis dimension," *IEEE Trans. Knowl. Data Eng.*, vol. 16, no. 4, pp. 385–395, 2004.
24. A. Bifet, G. Holmes, R. Kirkby, and B. Pfahringer, "MOA: Massive online analysis," *J. Mach. Learn. Res.*, vol. 11, pp. 1601–1604, 2010.
25. B. J. Kim and I. K. Kim, "Incremental nonlinear PCA for classification," in *Proc. Knowledge Discovery in Databases (PKDD 2004)*, Pisa, Italy, September 20–24, vol. 3202, pp. 291–300, 2004.
26. B. J. Kim, J. Y. Shim, C. H. Hwang, I. K. Kim, and J. H. Song, "Incremental feature extraction based on empirical kernel map," *Found. Intell. Sys.*, vol. 2871, pp. 440–444, 2003.
27. L. Hoegaerts, L. De Lathauwer, I. Goethals, J. A. K. Suykens, J. Vandewalle, and B. De Moor, "Efficiently updating and tracking the dominant kernel principal components," *Neural Netw.*, vol. 20, no. 2, pp. 220–229, 2007.
28. T. J. Chin and D. Suter. "Incremental kernel principal component analysis," *IEEE Trans. Image Process.*, vol. 16, no. 6, pp. 1662–1674, 2007.
29. X. Jiang, R. Snapp, Y. Motai, and X. Zhu, "Accelerated kernel feature analysis," in *Proceedings of IEEE Computer Society Conference on Computer Vision and Pattern Recognition*, New York, NY, 17–22 June, pp. 109–116, 2006.
30. Y. Motai and H. Yoshida, "Principal composite kernel feature analysis: Data-dependent kernel approach," *IEEE Trans. Knowl. Data Eng.*, vol. 25, no. 8, pp. 1863–1875, 2013.
31. H. Xiong, Y. Zhang, and X. W. Chen, "Data-dependent kernel machines for microarray data classification," *IEEE/ACM Trans. Comput. Biol. Bioinform.*, vol. 4, no. 4, pp. 583–595, 2007.
32. B. Schkopf and A. J. Smola. "Learning with kernels: Support vector machines, regularization, optimization, and beyond," *Adaptive computation and machine learning*. Cambridge, MA: MIT press, 2002.
33. H. Frhlich, O. Chapelle, and B. Scholkopf, "Feature selection for support vector machines by means of genetic algorithm," in *Proceedings of 15th IEEE International Conference on Tools with Artificial Intelligence*, Sacramento, California, November 3–5, pp. 142–148, 2003.

34. X. W. Chen, "Gene selection for cancer classification using bootstrapped genetic algorithms and support vector machines," in *Proc. IEEE International Conference on Computational Systems, Bioinformatics Conference*, Stanford, CA, August 11–14, pp. 504–505, 2003.
35. C. Park and S. B. Cho, "Genetic search for optimal ensemble of feature classifier pairs in DNA gene expression profiles," in *Proceedings of the International Joint Conference on Neural Networks*, vol. 3, pp. 1702–1707, 2003.
36. F. A. Sadjadi, "Polarimetric radar target classification using support vector machines", *Opt. Eng.*, vol. 47, no. 4, pp. 046201–046208, 2008.
37. B. Souza and A. de Carvalho, "Gene selection based on multi-class support vector machines and genetic algorithms," *Mol. Res.*, vol. 4, no. 3, pp. 599–607, 2005.
38. D. S. Paik, C. F. Beaulieu, G. D. Rubin, B. Acar, R. B. Jeffrey, Jr., J. Yee, J. Dey, and S. Napel, "Surface normal overlap: a computer-aided detection algorithm with application to colonic polyps and lung nodules in helical CT," *IEEE Trans. Med. Imaging*, vol. 23, pp. 661–75, 2004.
39. T. Damoulas and M. A. Girolami, "Probabilistic multi-class multi-kernel learning: on protein fold recognition and remote homology detection," *Bioinformatics*, vol. 24, no. 10, pp. 1264–1270, 2008.
40. S. Amari and S. Wu, "Improving support vector machine classifiers by modifying kernel functions," *Neural Netw.*, vol. 6, pp. 783–789, 1999.
41. A. K. Jerebko, J. D. Malley, M. Franaszek, and R. M. Summers, "Multiple neural network classification scheme for detection of colonic polyps in CT colonography data sets," *Acad. Radiol.*, vol. 10, pp. 154–60, 2003.
42. T. Briggs and T. Oates, "Discovering domain specific composite kernels," in *Proceedings of the 20'th National Conference on Artificial Intelligence*, Pittsburgh, Pennsylvania, July 9–13, pp. 732–738, 2005.
43. N. Cristianini, J. Kandola, A. Elisseeff, and J. Shawe-Taylor, "On kernel target alignment," in *Proceedings of. Neural Information Processing Systems*, Pittsburgh, Pennsylvania, July 9–13, pp. 367–373, 2005.
44. http://mldata.org/repository/data/viewslug/protein-fold-prediction-ucsd-mkl/
45. http://mldata.org/repository/data/viewslug/pendigits_ucsd_mkl/
46. http://archive.ics.uci.edu/ml/datasets/Internet+Advertisements. Accessed 20 November 2014.
47. A. Frank and A. Asuncion, *UCI machine learning repository*. Irvine, CA: University of California, School of Information and Computer Science, 2010. http://archive.ics.uci.edu/ml, Accessed 20 November 2014.
48. A. Rakotomamonjy, F. R. Bach, S. Canu, and Y. Grandvalet, "*Simple multiple kernel support vector machine*," *J. Mach. Learn. Res.*, vol. 9, pp. 2491–2521, 2008.
49. S. Qiu and T. Lane. "A framework for multiple kernel support vector regression and its applications to siRNA efficacy prediction," *IEEE/ACM Trans. Comput. Biol. Bioinform.*, vol. 6, no. 2, pp. 190–199, 2009.
50. N. Cristianini and J. Shawe-Taylor. *An introduction to support vector machines and other kernel-based learning methods*. Cambridge, England: Cambridge University Press, 2000.
51. M. Gönen and E. Alpaydin, "Localized multiple kernel learning," in *Proceedings of the 25th International Conference on Machine Learning*, Helsinki, Finland, July 5–9, 2008.

7

CONCLUSION

Kernel analysis (KA) has introduced how KA cooperates with different types of data formation. This book has addressed the difficulties that occur when trying to use KA in a biomedical environment with various scenarios. The proposed Principal Composite Kernel Analysis (PCKA) has been extended to provide the results of classification and prediction for computer-aided diagnosis (CAD). Data-variant KA was developed to address the need for a more efficient predictor of machine learning under the condition of given datasets such as offline, distributed, online, cloud, and longitudinal ones. The proposed new framework has provided a significant improvement in the classification performance without the use of additional KA methodologies. Table 7.1 summarizes how data-variant KA is designed for specific data types.

The advantages of each KA are listed, after the survey of other studies in Chapter 1, in the order of Chapter 2 to Chapter 6:

Offline KA

- Chose the appropriate kernels during the learning phase and adopted KA into the principle component techniques for the nonlinear data-space.
- Yielded higher detection performance of the accurate classification compared to the single kernel method, without taking more computational time across the variable data-size.

Group KA

- Applied offline learning algorithms to distributed databases. Networked multiple hospitals to access computed tomographic colonography (CTC), so that different types of cancer datasets can be accessed at any specific hospital. The number of true positive cases (cancer) in each hospital is usually small due to screening tests.

Data-Variant Kernel Analysis, First Edition. Yuichi Motai.

Table 7.1 Summary of framework data-variant KA methodology

Methods	Kernel formation	Data types
Offline KA	KA to choose the appropriate kernels as offline learning in training phase	Static data in a nonlinear learning feature space. Placed under the linear space via kernel tricks
Group KA	KA to group databases as a data-distributed extension of offline learning algorithms	Data extended into several databases. Each offline database is considered as a distributed database
Online KA	KA to update the feature space in an incremental or nonincremental manner	Up-to-date data. Updated database proceeds as more data is fed into the algorithm
Cloud KA	KA in the cloud network setting. Extending offline and online learning to access shared databases	Data stored in cloud servers. Distributed databases shared over public and private networks
Predictive KA	KA to predict future state. Prediction techniques using online learning	Longitudinal data. A time-transitional relationship from past to present

- Observed a larger amount of positive cases for big-data classification analysis. Showed comparable performance of speed, memory usage, and computational accuracy.

Online KA

- Proposed an online data association for the increasing amount of CTC databases. Online virtual colonography can be improved by exploiting heterogeneous information from larger databases of patients, diverse populations, and disparities.
- Achieved a high cancer classification performance in a realistic setting with a large number of patients.

Cloud KA

- Executed KA in the extension of distributed data in the cloud server hosting, where all the CTC data are shared from multiple hospital servers and executed in the cloud server.
- Provided equivalent classification results to other KA methods without additional, unnecessary, computational complexity.

Predictive KA

- Constructed a composite kernel for the anomaly data problem. We have developed nonstationary data associations to correspond with anomaly datasets.

- Yielded high classification and prediction accuracy in the change of cancer staging. The overall computing cost does not substantially change due to the light module of nonstationary prediction.

The project described in this book has resulted in the following three major contributions:

1. Development of extensive KA methods for different data configurations such as distributed and longitudinal.
2. Comparison of how several data-variant KAs accurately classify the cancer data.
3. Accomplishment of a high level of diagnosis in CTC by using more efficient KAs.

The result of this work is that the proposed data-variant KA will be usable in other applications with small additional investments in terms of resources, infrastructure, and operational costs. This technology is directly applicable to other areas besides simulation in the medical, consumer, and military fields. The representative MATLAB® codes are listed in the appendix as supplemental materials, so that readers can apply how the data-variant KA can be implemented.

APPENDIX A

The definitions of acronyms are shown in Table A.1, in the alphabetical order. The notations of mathematical equations are listed in Table A.2, in the order of chapters.

Table A.1 Acronyms definitions

1-SVM	One-class support vector machines
AKFA	Accelerated kernel feature analysis
AMD	Associated multiple databases
AOSVM	Adaptive one-class support vector machine
AR	Autoregressive
ARMA	Autoregressive moving average
AUC	Area under the curve
BKR	Budgeted kernel restructuring
BNN	Back-propagation neural network
CA	Constant acceleration
CAD	Computer-aided diagnosis/detection
CCA	Canonical correlation analysis
CT	Computed tomography
CTC	Computed tomographic colonography
CV	Constant velocity
DAG	Directed acrylic graph
DC	Difference of convex functions
DK	Diffusion kernels
DLBCL	Diffuse large B-cell lymphomas
DT	Decision tree
DTP	Discrete-time processes
EKF	Extended Kalman filter

(*continued*)

Data-Variant Kernel Analysis, First Edition. Yuichi Motai.

Table A.1 (Continued)

FA	Fuzzy ART
FDA	Fisher discriminant ratio
FL	Follicular lymphomas
FN	False negative
FP	False positive
FPR	False positive rate
FSM	Finite state model
GKFA	Group kernel feature analysis
GMKL	Generality multiple kernel learning
GRNN	General regression neural network
HBDA	Heterogeneous big data association
HD	Heterogeneous degree
HIPAA	Health Insurance Portability and Accountability Act
IKPCA	Incremental kernel principal component analysis
IMM	Interactive multiple model
KA	Kernel analysis
KAPA	Kernel affine projection algorithms
KARMA	Kernel autoregressive moving average
KCCA	Kernel canonical correlation analysis
KCD	Kernel change detection
KF	Kalman filter
KFA	Kernel feature analysis
KFDA	Kernel Fisher discriminant analysis
KKF	Kernel Kalman filter
KLMS	Kernel least mean square
KLT	Karhunen–Loève transform
KMA	Kernel multivariate analysis
K-NN	*K*-nearest neighbor
KOP	Kernel-based orthogonal projections
KOPLS	Kernel-based orthogonal projections to latent structures
KPCA	Kernel principal component analysis
KPLS	Kernel partial least squares regression
LACK	Longitudinal analysis of composite kernels
LCSS	Longest common subsequences
LDA	Linear discriminant analysis
MA	Moving average
MID	Medical image datasets
MK	Multiple kernel
MKL	Multiple kernel learning
MKSVM	Multiple kernel support vector machine
MRAC	Model reference adaptive control
MVA	Multivariate analysis
NCI	National cancer institute
NN	Neural network
NRMSE	Normalized root mean squared error
NSF	National science foundation
OPLS	Orthogonal projections to latent structures

Table A.1 (Continued)

PaaS	Platform as a service
PCA	Principal components analysis
PCKFA	Principal composite kernel feature analysis
PCR	Principal components regression
PE	Persistently exciting
PLS	Partial least squares
POD	Proper orthogonal decomposition
RBF	Radial basis function
RBF-NN	Radial basis function neural network
RKDA	Regularized kernel discriminant analysis
RKHS	Reproducing kernel Hilbert space
RMSE	Root mean square error
RNN	Radial bias function neural networks
ROC	Receiver operating characteristic
SKFA	Sparse kernel feature analysis
SKIPFSM	Separated kernel image processing using finite state machines
SMO	Sequential minimal optimization
SVM	Support vector machine
TLS	Transport layer security
TN	True negative
TP	True positive
TPR	True positive rate
TSVM	Twin support vector machine
VOI	Volume of interest

Table A.2 Symbol definitions

Chapter 1 Survey	
x	Input vector
$K(...)$	Kernel function
σ	Sparse of RBF/Gaussian function
β	Sigmoid kernel parameter(s)
$g(.)$	Riemannian metric
D	Number of base kernel
w	Non-negative weight
$\hat{x}$	Cross set($= x \times x$)
$g(.),\ h(.)$	Convex function
$D(.)$	Difference of convex function
N_i	Number of samples for class i
$K_{i,j}$	Kernel between ith and jth sample
F_λ^*	Fisher discriminant ratio with a positive regularization parameter λ
λ	Positive regularization parameter for the Fisher discriminant analysis
κ	Set of kernel functions K

(*continued*)

Table A.2 (Continued)

η	Output of neural network
w	Weight matrix for the neural network
α	Lagrange multiplier
$L(.)$	Lagrangian
y, y_i	Output class label
$\Phi(.)$	Mapping function from input space to Hilbert space
H	Hilbert Space
$R, \Re$	Real number space
tr(.)	Trace of matrix inside the parenthesis
$e(.)$	Error function
F	Transition matrix
B	Control-input matrix
H	Measurement matrix
$u(t)$	n-Dimensional known vector
$z(t)$	Measurement vector
W	Zero-mean white Gaussian process noise
V	Zero-mean white Gaussian measurement noise
$E[.]$	Expected value
$\hat{x}$	Predicted position of x
Chapter 2 Offline Kernel Analysis	
S	Scatter matrix
e_j	jth eigenvector of S
λ_j	jth eigenvalue of S
x_i	Input data
$\Phi(.)$	Mapping function from input space to Hilbert space
n	Number of samples
K	Gram matrix
l	Number of principal components
S_{bi}	Between class scatter matrix
S_{wi}	Within class scatter matrix
$A(\cdot, \cdot)$	Alignment factor
y_i	Output class label
k_{ij}	Element of Gram matrix
J	Fisher scalar
r	Number of class
λ_λ	Eigenvalue of Fisher scalar
λ_*	Largest eigenvalue
ρ	Composite coefficient
δ	Eigenvalues of kernel alignment
$\hat{\rho}$	Optimum composite coefficient
$\lVert \cdot \rVert_F$	Frobeneus norm
$<\cdot, \cdot>_F$	Frobeneus inner product
Chapter 3 Group Kernel Feature Analysis	
$q_l(.)$	Factor function
L	Base kernel label
$p_s = (\cdot, \cdot)$	Base kernel

Table A.2 (*Continued*)

α_{lm}	Combination coefficient
K_l	Data-dependent kernel
Q_l	Factor function matrix
Q_l'	Updated Q_l atrix
P_l	Base kernel matrix
P_l'	Updated P_l matrix
J	Fisher scalar
λ_l	Eigenvalue of Fisher scalar
λ_*	Largest eigenvalue
ξ	Ratio of the class separability
rf	Residue factor
n_r	Number of data for the rth cluster
$K^s_{com}(\rho)$	Composite kernel
s	One of six composite kernels
$A(k_1, k_2)$	Empirical alignment between kernels k_1 and k_2
K^s_g	Group kernel
δ_d	Eigenvalues of the database d
$J'_*(a'_l)$	Class separability yielded by the most dominant kernel for dataset (subsets) of database
$J_*(a'_l))$	Class separability yielded by the most dominant kernel for the entire database
a'_*	Combination coefficients of the most dominant kernel among the subsets
$\overrightarrow{a}_*$	Mean of combination coefficients of all databases
Chapter 4 Online Kernel Analysis	
p_r	One of the chosen kernels
$K_{comp}(\rho)$	Optimized combination of single kernels—composite kernel
$J(\rho)$	Generalized Raleigh coefficient
ξ	Ratio of class separability of composite online data and offline data
λ_*'	Most dominant eigenvalue of composite data
λ_*	Most dominant eigenvalue of the four different kernels for the offline data
η	Threshold value for class separability ratio, ξ
$J'_*(\alpha'_r)$	Class separability yielded by the most dominant kernel for composite data
$J_*(\alpha_r)$	Class separability yielded by the most dominant kernel for offline data
$K_n^{r'}(\rho)$	Updated Gram matrix
HD	Heterogeneous degree
α'_*	Maximum combination coefficient of four different kernels
$\overline{\alpha_*}$	Mean of combination coefficients
$\overline{\alpha'_*}$	Most dominant kernel among the kernels available
$K^{n'}{}_{comp}(\rho)$	Optimized composite kernel for the new composite data

(*continued*)

Table A.2 (*Continued*)

${}^{t+1}(K^{ri}_{comp})_t$	Update of Gram matrix from time t to $t+1$
${}^{t+1}(\mathrm{Frob})_t$	Update of Gram matrix from time t to $t+1$
${}^{(t+1)}(\lambda_*{}')_t$	Largest eigenvalue of the data (of the dominant kernel) received from time t to $t+1$
η'	Threshold value of ${}^{(t+1)}(\xi)_t$
Chapter 5 Cloud Kernel Analysis	
None	
Chapter 6 Predictive Kernel Analysis	
$p_s = (\cdot,\cdot)$	Base kernel
α_l	Combination coefficient
Q_l	Factor function matrix
$Q'{}_l$	Updated Q_l matrix
P_l	Base kernel matrix
P'_l	The updated P_l matrix
λ_l	Eigenvalue
λ^*_l	Largest eigenvalue
$K^s_{com}(\rho)$	Composite kernel
s	One of six composite kernel
δ_d	Eigenvalues of the database d

APPENDIX B

REPRESENTATIVE MATLAB CODES

The source codes are implemented in the platform of MATLAB® 7.10.0, R2010a(*). These codes developed are built on top of Statistical Pattern Recognition Toolbox Version 2.10(**).

*http://www.mathworks.com.

**http://cmp.felk.cvut.cz/cmp/software/stprtool/.

The M-codes in Appendix B include some comments, and extract the core parts of M-codes, representing the corresponding algorithms in the book.

The Chapter 2 source codes for offline kernel analysis are as follows:

B.1 Accelerated Kernel Feature Analysis (AKFA.m)
B.2 Experimental Evaluations (cross_validation.m)

The Chapter 3 source codes for group kernel analysis are as follows:

B.3 Group Kernel Analysis (multi_kernel.m)

The Chapter 4 source codes for online kernel analysis are as follows:

B.4 Online Composite Kernel Analysis (PCKFA.m)
B.5 Online Data Sequences Contol (LTSTSM.m)
B.6 Alignment Factor (AF_fig.m)

Data-Variant Kernel Analysis, First Edition. Yuichi Motai.

The Chapter 5 source codes for cloud kernel analysis are as follows:

B.7 Cloud Kernel Analysis (DTldr.m)
B.8 Plot Computation Time (cputime_plot.m)
B.9 Parallelization (kernel_par.m)

B.1 ACCELERATED KERNEL FEATURE ANALYSIS

```
function model=AKFA(X,options,delta)
% This function implements the Accelerated Kernel Fea-
ture Analysis
% algorithm.
% X is the input data
% options contain the all(except delta size) parame-
ters of the algorithm
% delta is the threshold value of kernel


num_data = size(X,2);
I=options.new_dim;
C=zeros(I,I);

start_time = cputime;  % for calculating the
                         computational time

K=kernel(X,options.ker,options.arg); % Gram Matrix
J = ones(num_data,num_data)/num_data;
K = K - J*K - K*J + J*K*J; % Centering matrix

tempK=K; % tempK is the updated Gram Matrix

idx=zeros(1,I);
mark=ones(1,num_data); % =0 if i-th data selected as
                          feature or deleted

for i=1:I
   if i==1
      c=sum(tempK.^2)./diag(tempK)';
   else
      c = zeros(1,num_data);
      for j=1:num_data
         if mark(j)>0
            if tempK(j,j)>delta
               c(j)=sum((tempK(j,:).*mark).^2)/tempK(j,j);
            else
```

```
                mark(j)=0;
            end
        end
    end
end

[~,idx(i)]=max(c);
mark(idx(i))=0;
for j=1:num_data % update the Gram Matrix
    if mark(j)>0
        tempK(j,:)=tempK(j,:)-tempK(j,idx(i))*tempK(idx(i),:)
        .*mark/(tempK(idx(i),idx(i)));
    end
end

C(i,i) = 1 / sqrt(tempK(idx(i),idx(i)));
C(i,1:(i-1))=-sum(K(idx(i),idx(1:(i-1)))*C(1:(i-1),1:
(i-1))'*C(1:(i-1),1:(i-1)),1)*C(i,i);
end

end_time = cputime-start_time;
model.cputime = end_time;
% output structure
Alpha=zeros(num_data,I);
for i=1:I
    for j=1:I
        Alpha(idx(j),i)=C(i,j);
    end
end
model.Alpha=(eye(num_data,num_data)-J)*Alpha;

Jt=ones(num_data,1)/num_data;
model.b = Alpha'*(J*K*Jt-K*Jt);

model.sv.X = X;
model.options = options;
model.nsv = num_data;
model.coefficient=C;

Error=zeros(1,num_data);
for i=1:num_data
     Error(i)=K(i,i)-K(i,idx(1:I))*(C'*C)*K(idx(1:I),i);
end

model.mse=sum(Error)/num_data;
return;
```

B.2 EXPERIMENTAL EVALUATIONS

```
function [accuracy,confusion]=cross_validation(method,options,
trndata,tstdata,FP_train_size,datacount,num_neighbor)

% This function contains sample control process of
 experiments using
% Principal Composite Kernel Feature Analysis Algorithm.
 Support Vector
% Machine is employed for the final classification

accuracy=zeros(1,num_neighbor);
confusion=zeros(2,2);
%%%%%%%%%%%%%%%%%%% single kernel method %%%%%%%%%%%%%%%%%%%%%
switch method
    case 1
        model=kpca(trndata.X,options);  % KPCA
    case 2
        model=AKFA(trndata.X,options,datacount,0.0);  % AKFA
                                                      or ACKFA
    case 3
        model=OAKFA(trndata.X,options,FP_train_size,datacount,
         0.0);  % Online AKFA
    case 4
        model=SIGSKFA4(trndata.X,options);  % SKFA
end
%%%%%%%%%%%%%%%%%%%%%%%%%%%%%%%%%%%%%%%%%%%%%%%%%%%%%%%%%%%%%%
%%%%%%%%%%%%%%%%%%% composite kernel method %%%%%%%%%%%%%%%%%%
% switch method
%     case 1
%         model=MyKPCA(trndata.X,options);  % KPCA
%     case 2
%         model=MyAKFA(trndata.X,options,0.1);  % AKFA
          or ACKFA
%         model=MyAKFA(trndata.X,options,0.0);  % AKFA
          or ACKFA
%     case 3
%         model=OAKFA(trndata.X,options,FP_train_size,
datacount,0.0);  % Online AKFA
%     case 4
%         model=COMSKFA4(trndata.X,options);  % SKFA
% end
%%%%%%%%%%%%%%%%%%%%%%%%%%%%%%%%%%%%%%%%%%%%%%%%%%%%%%%%%%%%%%


trnreduced = kernelproj(trndata.X,model);
tstreduced = kernelproj(tstdata.X,model);
```

```
% % Projtrnreduced = kernelproj(trndata.X,model); % Training
 data was projected into the reduced dimensions
% % ProjVar = var(Projtrnreduced,0,2); % Variance
 along each new dimension
% % load ProjV
% % ProjVariance(:,3) = ProjVar;
% % save('ProjV.mat','ProjVariance');
% %
% % figure; plot(ProjVar);
% % title('Variance along each selected dimension');
% % xlabel('New Dimension'); ylabel('Variance');
% % params = 'RBF', 0.1, 'Perceptron', 0.5;
%
% [test_targets, a_star] = SVM(trnreduced, trndata.y, tstre-
duced, 'Poly', 0.01, 'Perceptron', 0.1)
%
%
% ypred=test_targets-tstdata.y;
% svm_accuracy=length(find(ypred==0))/length(test_targets)
%
%
% %%%%%%%% one as test the other serve as train serials %%%%%
% % B.X=trnreduced;      %copy data.X and data.y to B.X and B.y
 for processing
% % B.y=trndata.y;
% % accuracy=0;
% % wc=1;
% % for i=1:trndata.size       %use leave-one-out method.
% %     temp=B.X(:,1);      %In each cycle, switch the test
 data into B.X (:,1),
% %     B.X(:,1)=B.X(:,i);
% %     B.X(:,i)=temp;
% %     temp_label=B.y(1); %and switch the label of the test
 data to B.y(1)
% %     B.y(1)=B.y(i);
% %     B.y(i)=temp_label;
% %     trndata.X=B.X(:,2:trndata.size);  % use all the
          left data as training data
% %     trndata.y=B.y(2:trndata.size);
% %     tstdata.X=B.X(:,1);
% %     tstdata.y=B.y(1);
% %     [AlphaY, SVs, Bias, Parameters, nSV, nLabel] = SVM-
Train(trndata.X, trndata.y);
% %     [Labels, DecisionValue]= SVMClass(tstdata.X, AlphaY,
          SVs, Bias, Parameters, nSV, tstdata.y);
% % %     model = svm2(trndata,options );  %svm2 () is to
 train a binary SVM classifier
% % %     ypred = svmclass( tstdata.X, model ); % svmclass()
 use the trained model to classify the test data
```

```
% %     if ypred==tstdata.y
% %         accuracy=accuracy+1;
% %     else
% %         wrong(wc)=i;   % record the index of
 wrong-classified data
% %         wc=wc+1;
% %     end
% % end
% % accuracy=accuracy/trndata.size  % calculate the accu-
racy percentage
% %%%%%%%%%%%%%%%%%%%%%%%%%%%%%%%%%%%%%%%%%%%%%%%%%%%%%%%%%%%%

%%%%%%%%%%%%%%%%%%%%%% KNN classifier %%%%%%%%%%%%%%%%%%%%%
for k=1:num_neighbor
  TPs(k) = 0;
  TNs(k) = 0;
  FNs(k) = 0;
  FPs(k) = 0;
  sensitivity(k) = 0;
  specificity(k) = 0;
  accuracy(k) = 0;
  classification_accuracy(k) = 0;
end

wrong=zeros(num_neighbor,tstdata.size);

for i=1:tstdata.size
    tstreduced=kernelproj(tstdata.X(:,i),model);
    tmp1= sum((trnreduced-(tstreduced*ones(1,(trndata.size))))
    .^2,1);
    [y,id1]=sort(tmp1);

    for k=1:num_neighbor
        nnlabels=trndata.y(id1(1:k));
        for j=1:2
            countlabels(j)=length(find(nnlabels==(j-1)));
        end
        [y,id2]=max(countlabels);  % Check the mostly selected
         cluster, id2 can be 1 or 2

        if id2==(tstdata.y(i)+1)
            accuracy(k)=accuracy(k)+1;
            if i>tstdata.FPsize
                TPs(k) = TPs(k) + 1;
            else
                TNs(k) = TNs(k) + 1;
            end
        else
            wrong(k,i)=1;  % Incorrectly classified
```

```
            confusion(tstdata.y(i)+1,id2) =
             confusion(tstdata.y(i)+1,id2) + 1;
            if i>tstdata.FPsize
                FNs(k) = FNs(k) + 1;
            else
                FPs(k) = FPs(k) + 1;
            end
        end  % end of (id2==(tstdata.y(i)+1))
     end  % end of (k=1:num_neighbor)
end  % end of (i=1:tstdata.size)

for k=1:num_neighbor
  sensitivity(k) = TPs(k)/(TPs(k) + FNs(k));
  specificity(k) = TNs(k)/(TNs(k) + FPs(k));
  classification_accuracy(k) = (TPs(k) + TNs(k))/(TPs(k) +
   TNs(k) + FPs(k) + FNs(k));
end

accuracy = accuracy/tstdata.size
%%%%%%%%%%%%%%%%%%%%%%%%%%%%%%%%%%%%%%%%%%%%%%%%%%%%%%%%%%%%%%%
return;

% save('wrong.mat','wrong');
% figure;   % Plots for each iteration of the datacount value
% plot(accuracy,'g-');
% hold on
% plot(sensitivity,'r-');
% plot(specificity,'b-');
% legend('Accuracy', 'Sensitivity', 'Specificity');
% % title('Accuracy, Sensitivity and Specificity vs.
    Number of Neighbors');
% xlabel('Number of Neighbors'); % Plot against num_neighbor
                                   values
% hold off
```

B.3 GROUP KERNEL ANALYSIS

```
function Ker = multi_kernel(options, test_patterns,
 train_patterns, Npp, Nnn)
% GKFA multi kernel optimizer.

Np = Nnn;
Nn = Npp;
if Np == 0
    Np = 1;
    Nn = Nn-1;
end
if Nn == 0
```

```
    Np = Np-1;
    Nn = 1;
end
NN = Np+Nn;
Y = test_patterns;
X = train_patterns;
% gets dimensions
[dim,num_data] = size(X);

% process input arguments
%- - - - - - - - - - - - - - - - - - - - - - - - - -
if nargin > 2, options = []; else options=c2s(options); end
if ~isfield(options,'ker'), options.ker = 'linear'; end
if ~isfield(options,'arg'), options.arg = 1; end
if ~isfield(options,'new_dim'), options.new_dim = dim; end

% compute kernel matrix
%K = kernel(X,options.ker,options.arg);
%%%%%%%%%%%%%%%%%%%%%%%%%%%%%%%%%%%%%%%%%%%%%


bk1 = kernel(Y, X, options.ker1,options.arg1); % Normalization

bk2 = kernel(Y, X, options.ker2,[options.arg2 options.arg3]);

for i = 1:NN
    for j= 1:NN
%         if bk2(i,j)<-0.5;
        if bk2(i,j)<0;
            bk2(i,j)=-1*bk2(i,j);
        else bk2(i,j) = bk2(i,j);
        end
        bk2(i,j) = bk2(i,j)/(10^3);
    end
end


%%%%%%%%%%%%%%%%%%%%%%%%%%%%%%%%%%%%%%%%%%%%%%%%%%%%%%%%%%%%%%%%%%%%%%%%
bk3 = kernel(Y, X, options.ker3,options.arg4);
d = ones(length(bk3),1);

k1 = [d,bk3];
% now creating mo,no, bo,wo matrices...;

A11 = 1/Np*bk1(1:Np,1:Np);
A22 = 1/Nn*bk1(Np+1:NN,Np+1:NN);
a = zeros(Np,Nn);
b = zeros(Nn,Np);
B1 = [A11 a; b A22] - 1/NN*bk1;
```

```
%B1 = (B1-ones(length(B1),1)*mean(B1))./(ones(length(B1),1)
*std(B1));
W1 = diag(diag(bk1)) - [A11 a; b A22];
%W1 = (W1-ones(length(W1),1)*mean(W1))./(ones(length(W1),1)
*std(W1));

D11= 1/Np*bk2(1:Np,1:Np);
D22= 1/Nn*bk2(Np+1:NN,Np+1:NN);
B2 = [D11 a; b D22] - 1/NN*bk2;
%B2 = (B2-ones(length(B2),1)*mean(B2))./(ones(length(B2),1)
*std(B2));

W2 = diag(diag(bk2)) - [D11 a; b D22];
%W2 = (B2-ones(length(B2),1)*mean(B2))./(ones(length(B2),1)
*std(B2));


M1 = k1'*B1*k1;
 M1 = (M1-ones(length(M1),1)*mean(M1))./(ones(length(M1),1)
*std(M1));

N1 = k1'*W1*k1;%+0.1* eye(length(M1));
 N1 = (N1-ones(length(N1),1)*mean(N1))./(ones(length(N1),1)
*std(N1));

M2 = k1'*B2*k1;
 M2 = (M2-ones(length(M2),1)*mean(M2))./(ones(length(M2),1)
*std(M2));

N2 = k1'*W2*k1;%+0.1* eye(length(M2));
N2 = (N2-ones(length(N2),1)*mean(N2))./(ones(length(N2),1)
*std(N2));


[alp1,lambda1]= eig(M1,N1);
[c1,i1] = max(diag(lambda1));
AL1 = alp1(i1,:)';
[alp2,lambda2]= eig(M2,N2);
[c2,i2] = max(diag(lambda2));
AL2 = alp2(i2,:)';
%maximal eigenvalue
% c1
% c2
%%%%%%%%%%%%%%%%%%%%%%%%%%%%%%%%%%%%%%%%%%%%%%%%%%%%%%%%%%%%%%%%%%%%%%
choose the kernels corresponding to most dominant eigenvalues
%%%%%%%%%%%%%%%%%%%%%%%%%%%%%%%%%%%%%%%%%%%%%%%%%%%%%%%%%%%%%%%%%%%%%%
AL1 = [1 zeros(1,NN)]';
AL2 = [1 zeros(1,NN)]';
N_t = 100;
```

```
for t = 1:N_t


%     j11= AL1'*M1*AL1;
%     j12= AL1'*N1*AL1;
%     j1 = j11/j12;
%     j21= AL2'*M2*AL2;
%     j22= AL2'*N2*AL2;
%     j2 = j21/j22;
    q1 = k1*AL1;
    q2 = k1*AL2;
    j11= q1'*B1*q1;
    j12= q1'*W1*q1;
    j1 = j11/j12;
    j21= q2'*B2*q2;
    j22= q2'*W2*q2;
    j2 = j21/j22;
    % not using the decreasing rate for eta.
    tt = 0.01*(1-t/N_t);
    tt = 0.005*(1-t/N_t);

    AL1= AL1+ tt*(1/j12*M1 - j1/j12*N1)*AL1;%for gaussian
                                             kernel;
    AL2= AL2 + tt*(1/j22*M2 - j2/j22*N2)*AL2;%for polynomial
                                              kernel;

    AL1 = (AL1-ones(length(AL1),1)*mean(AL1))./(ones(length
(AL1),1)*std(AL1));
    AL2 = (AL2-ones(length(AL1),1)*mean(AL2))./(ones(length
(AL1),1)*std(AL2));

%    normalization of all values.
%     trndata.X=(trndata.X-ones(dim,1)*mean(trndata.X))./
(ones(dim,1)*std(trndata.X));
  %%%%%%normalize the value of Alpha;
end

%dont worry about calculating q(x) and q(y) beacause K = QK0Q
 and Q is
%nothing but the diagonal matrix og q. so just find q and q is
 nothing but
%k1*alpha..so concentrate on finding alpha..and Q = diag(q)..

% j1
% j2

Q1= diag(q1);
Q2= diag(q2);
K1 =  Q1*bk1*Q1;
K2 = Q2*bk2*Q2;
```

```
%%%%%%%%%%%%%%%%%%%%%%%%%%%%%%%%%%%%%%%%%%%%%%%%%%%%%%%%%%%%%%%
              calculate the most dominant kernels
%%%%%%%%%%%%%%%%%%%%%%%%%%%%%%%%%%%%%%%%%%%%%%%%%%%%%%%%%%%%%%%
%part b:
y=[zeros(1,Np) ones(1,Nn)]';              %target vector
y1= y*y';%this is the correct definition...
     the matrix yy' should be order nxn
% this is the correct definition of inner prod-
uct,cross checked it.
u1 = trace(K1'*y1);
u2 = trace(K2'*y1); % this is the correct definition of inner
%product,cross checked it.

U1 = [u1*u1 u1*u2; u2*u1 u2*u2];

V = [trace(K1'*K1) trace(K1'*K2);trace(K2'*K1) trace(K2'*K2)];

% If the matrix V is singular then we will use eig(U,V) to
 compute the
% eigen values and eigen vectors. Otherwise we simply use
 eig(inv(V)*U) to
% compute the eigen vectors and values respectively because
 now the generalized eigen problem
% is reduced to a standard eigen value problem

%Checking whether the matrix V is singular or not.
%1)If its rank is equal toits order then it is a non
 singular matrix.
%2) its determinent should be a non zero value.

order = max(size(V));

if rank(V)==order&& det(V)~=0
   [rho,lambda3]= eig(inv(V)*U1);
else
   [rho,lambda3]= eig(U1,V);
end

%rho = rho';
[~,i3] = max(diag(lambda3));

reqrho = rho(i3,:)';
s = exp(reqrho(1,1))+ exp(reqrho(1,2));

if reqrho(1,1)<0 && reqrho(1,2)<0
    reqrho1 = exp(reqrho)./s;
else if reqrho(1,1)>0 && reqrho(1,2)<0
    reqrho1 = [reqrho(1,1) exp(reqrho(1,2))/s];
    else if reqrho(1,1)<0 && reqrho(1,2)>0
             reqrho1 = [exp(reqrho(1,1))/s reqrho(1,2)];
```

```
        else reqrho1 = exp(reqrho)./s;
        end
    end
end
% reqrho1
%[rho,lambda2]= eig(U1,V)
%[r1,l1]= eig(inv(V)*U1)
r1 = reqrho1(1,1);
r2 = reqrho1(1,2);

%K = bk1;
Ker = 0.9852*K1+0.1527*K2;%+0.12*K2;%+0.1192*K2;
return;
```

B.4 ONLINE COMPOSITE KERNEL ANALYSIS

```
function [K_comp,rho,r_sel]=PCKFA(X,op)
% this function is for kernel selection, combination
% optimization. It employs the PC-KFA algorithm

%% edit this cell to modify your input (for debug purposes)
% % loader
% load 'FP.mat'
% load 'TP.mat'
% X=[TP,FP];

% Sample Parameters
% op.ker{1}='rbf'; op.arg{1}=32;
% op.ker{3}='linear'; op.arg{3}=[.5,1.5];
% % op.ker{2}='poly'; op.arg{2}=2;
% op.ker{2}='sigmoid'; op.arg{2}=32;
% op.n1=size(TP,2); op.n2=size(FP,2);

%%


n=size(X,2);
if ~isa(X,'double')
    X=double(X);
end

% 3.1 kernel selection
if n~=op.n1+op.n2
    error('n~=op.n1+op.n2');
end
P=cell(length(op.ker),1);
P11=P;P22=P;B0=P;W0=P;M0=P;N0=P;Q=P;K=P;%K0=P;
% alpha_star=zeros(length(op.ker),1);lambda_star=alpha_star;
```

```
J = ones(n,n)/n; % centering

for r=1:length(op.ker)
    P{r} = kernel(X,op.ker{r},op.arg{r});
    P{r} = P{r} - J*P{r} - P{r}*J + J*P{r}*J; % centering
    P11{r}=P{r}(1:op.n1,1:op.n1);
    P22{r}=P{r}(op.n1+1:end,op.n1+1:end);

    B0{r}=[P11{r}/op.n1 zeros(op.n1,op.n2);zeros(op.n2,op.n1)
 P22{r}/op.n2];
    W0{r}=diag(diag(P{r}))-B0{r};
    B0{r}=B0{r}-P{r}/n;
    if strcmp(op.ker{r},'rbf')
        Ktemp=kernel(X,op.ker{r},16);% 16??
        Ktemp = Ktemp - J*Ktemp - Ktemp*J + J*Ktemp*J;
 % centering
        K0=[ones(n,1) Ktemp];
% K0{r}=[ones(n,1) P{r}];
    end
end
for r=1:length(op.ker)
    M0{r}=K0'*B0{r}*K0;
    N0{r}=K0'*W0{r}*K0;
    [alpha_star(:,r),lambda_star(r)]=eigs(M0{r},N0{r},1);
%     alpha_star(:,r)=t1;
end

if ~exist('op.p','var')
    op.p=2;
end
% r_sel=zeros(op.p,1);
[~,r_sel]=sort(lambda_star,'descend');
for t1=1:op.p
    [~,r_sel(t1)]=find(lambda_star==t0(t1));
end
for r=1:op.p
    q(:,r)=K0*alpha_star(:,r_sel(r));
    Q{r}=diag(q(:,r));
    K{r}=Q{r}*P{r_sel(r)}*Q{r};
end

% clear P B0 W0 M0 N0 P11 P22 Q
% kernel combinatory optimization

yyT=[ones(op.n1),-ones(op.n1,op.n2);...
    -ones(op.n2,op.n1),ones(op.n2)];
for t0=1:op.p

    for t1=1:op.p%t0
```

```
        ui=(sum(sum(K{t0}.*yyT)))^.5;
        uj=(sum(sum(K{t1}.*yyT)))^.5;
        U(t0,t1)=ui*uj;
        V(t0,t1)=(sum(sum(K{t0}.*K{t1})))^.5;
    end
end
[rho,D]=eigs(U,V,1);
% [rho,~]=svd(U,V,1);

K_comp=zeros(n);
for t0=1:op.p
    K_comp=K_comp+K{r}*rho(t0)^2;
end
```

B.5 ONLINE DATA SEQUENCES CONTOL

```
function LTSTSM(don)
% this function controls heterogeneous/homogeneous check of
 online data.
% don is the online data packet

global doff

t1=don.K;
t0=doff.K;
zet1=don.op.arg{1};
zet2=doff.op.arg{1};

while t1>t0

    if zet1>zet2
        don1=don;
        sz=round(size(don.K,1)/2);

        if sz<doff.thrsh
            t2=size(don.Xs,2)+1;
            doff.Xs=[doff.Xs don.Xs];
            doff.K(end-t2:end,end-t2:end)=[doff.K don.K];
            return
        end

        don1.K=don.K(1:sz,1:sz);
        don.K=don.K(1+sz:end,1+sz:end);
        LTSTSM(don);
        LTSTSM(don1);

    else
```

```
        t2=size(don.Xs,2)+1;
        doff.Xs(end-t2:end)=don.Xs;
        doff.K(end-t2:end,end-t2:end)=don.K;
    end

end

end
```

B.6 ALIGNMENT FACTOR

```
% Sample script for plotting the Alignment factor as shown
 in the book

load data_AF.mat

% plot(per(1:length(m1)),m1,'r','Linewidth',1.5)
plot(per(1:length(m1)),m1','r.-','Linewidth',1.5,
 'MarkerSize',20)
% plot(per,m1','r')
hold on
plot(per(1:length(m2))+sft,m2','g.-','Linewidth',1.5,
 'MarkerSize',20)
hold on
plot(per(1:length(m3))+2*sft,m3','b.-','Linewidth',1.5,
 'MarkerSize',20)
legend('Large Online data','Medium Online data','Small Online
 data','Location','SouthEast')
% legend('Small training set(<7% of available data)','Moderate
 training set(<12% of available data)','Large training set
(<16% of available data)')
hold on

% plot(per(1:length(s1)),m1'+s1'/3,'r:');hold on
for k=1:length(s1)
    if m1(k)+s1(k)/2>.99
        y=[m1(k)-s1(k)/2 .99];
    else
        y=[m1(k)-s1(k)/2 m1(k)+s1(k)/2];
    end
    plot(per(k)*[1 1],y,'rs:', 'MarkerSize',2,
'MarkerFaceColor','r');hold on
end
for k=1:length(s2)
    plot(per(k)*[1 1]+sft,[m2(k)-s2(k)/2.2 m2(k)+s2(k)/2.2],
'gs:', 'MarkerSize',2,'MarkerFaceColor','g');hold on
end
```

```
for k=1:length(s3)
    plot(per(k)*[1 1]+sft*2,[m3(k)-s3(k)/2.5 m3(k)+s3(k)/2.5],
'bs:', 'MarkerSize',2,'MarkerFaceColor','b');hold on
end


xlabel('Ratio of online data and offline data');
ylabel('Alignment Factor (AF)');
axis([0 9.1 0.25 1.0])
% title('Alignment Factor for different amount of off-line
 training')

% legend('type 1-(offline training 0.5%)','type 2-(offline
 training 5%)','type 3-(offline training 10%)')
```

B.7 CLOUD KERNEL ANALYSIS

```
%% Compute large Data Gram Matrix
% this sample computes sigmoid kernel

dt_loader=matfile('....\DT_normal.mat')
idx=round((1:10)'*num3749/10);
idx=[[1;idx(1:9)+1] idx];
Ksigmoid(num3749,num3749)=0;
datestr(now)
t3=cputime;
t2o=0;t2n=1;
seq1=[];
for t1=1:size(idx,1)
    dt1=dt_loader.DT(:,idx(t1,1):idx(t1,2));
    for tc=1:t1
        if tc~=1
            t2n=t2n+1;
            if t2n>t1
                t2n=1;
            end
        end
        if t2o~=t2n
            dt2=dt_loader.DT(:,idx(t2n,1):idx(t2n,2));
            t2o=t2n;
        end
        if t1==t2n
            Ksigmoid(idx(t1,1):idx(t1,2),idx(t2n,1):
idx(t2n,2))=kernel(dt1,'sigmoid',1e-5);
        else
            Ksigmoid(idx(t1,1):idx(t1,2),idx(t2n,1):
idx(t2n,2))=kernel(dt1,dt2,'sigmoid',1e-5);
```

```
            Ksigmoid(idx(t2n,1):idx(t2n,2),idx(t1,1):
idx(t1,2))=Ksigmoid(idx(t1,1):idx(t1,2),idx(t2n,1):
idx(t2n,2))';
        end
        seq1=[seq1;t1 t2n tc];
    end
end
t3=cputime-t3
datestr(now)
```

B.8 PLOT COMPUTATION TIME

```
%% Plot computation time
% load cputime_results.mat

for t1=1:length(sz1)
    sz1(t1)=length(idx{t1});
    trn.X=Krbf(idx{t1},idx{t1});
    trn.y=2*ones(1,sz1(t1));
    idx2=find(idx{t1}<137);
    trn.y(idx2)=1;
    for t2=1:length(nd)
        t_tr(t1,t2)=cputime;
        model=lda(trn,nd(t2));
        t_tr(t1,t2)=cputime-t_tr(t1,t2);
    end
    disp(t1)
end

%%
figure(8), clf
sf=10;%scale factor
plot(szdesk',t_trdesk(:,1),'r-.','LineWidth',1.5); hold on
plot(szdesk',t_trdesk(:,2),'b-.','LineWidth',1.5); hold on
plot(szdesk',t_trdesk(:,3),'g-.','LineWidth',1.5); hold on

plot(sz1(1:end-3),t_tr((1:end-3),1)/sf,'r','LineWidth',1.5);
 hold on
plot(sz1(1:end-3),t_tr((1:end-3),2)/sf,'b','LineWidth',1.5);
 hold on
plot(sz1(1:end-3),t_tr((1:end-3),3)/sf,'g','LineWidth',1.5);
 hold on

title('Training Time from Gram Matrix')
xlabel('Number of Data')
ylabel('Total Computation Time (second)')
```

```
legend('Database 1: Desktop','Database 2: Desktop','
        Database 3: Desktop',...
    'Database 1: Private Cloud','Database 2: Pri-
vate Cloud','Database 2: Private Cloud',...
    'Location','NorthWest')
text(40,5,['Cloud Time Scaled Down by Fac-
tor of ' num2str(sf)])

% figure(9), clf
% plot(sz1(1:end-1),t_tr((1:end-1),1),'r','LineWidth',1.5);
 hold on
% plot(sz1(1:end-1),t_tr((1:end-1),2),'b','LineWidth',1.5);
 hold on
% plot(sz1(1:end-1),t_tr((1:end-1),3),'g','LineWidth',1.5);
 hold on
```

B.9 PARALLELIZATION

```
%function K = kernel_par(X,ker,arg,N_WORKER)
% kernel matrix computation, large dimension parallel
 processing method
%
% Synopsis:
%  K = kernel(X,ker,arg,N_WORKER)
%
% Description:
%  K = kernel( X, ker, arg ) returns kernel matrix K [n x n]
%
%    K(i,j) = k(X(:,i),X(:,j))  for all i=1..n, j=1..n,
%
%   where k: a x b -> R is a kernel function given by
%   identifier ker and argument arg:
%
% Identifier   Name           Definition
% 'linear'...  linear kernel  k(a,b) = (a'*b)*arg
% 'poly'   ... polynomial     k(a,b) = (a'*b/arg(2)+1)^arg(1)
% 'rbf'    ... RBF            k(a,b) = exp(-0.5*||a-b||^2/arg^2)
%             (Gaussian)
% 'sigmoid'... Sigmoidal      k(a,b) = tanh(arg(1)*(a'*b)+arg(2))
%
% Input:
%  X [dim x n] Single matrix of input vectors.
%  ker [string] Kernel identifier.
%  arg [1 x  narg] Kernel argument.
%  N_WORKER [1x1] Number of divisions to be used.
%  (dim/N_WORKER) must be an integer
%
```

```
% Output:
%  K [n × n] Kernel matrix.
%
% Example:
%  ppl=parpool(2)
%  X = rand(400,50);
%  K = kernel_par( X, 'rbf', 1);
%  figure; pcolor( K );
%  delete(ppl)
%
% Script written by Nahian Alam Siddique, 2014

[dim,n]=size(X);
if mod(dim,N_WORKER)
    disp('dim must be divisible by N_WORKER')
    K=-1;
    return;
else
    X=reshape(X',[n,dim/N_WORKER,N_WORKER]);
end

if ~iscell(ker) % single kernel
    % if dim>n
    Ktemp=zeros(n,n,N_WORKER);

    switch ker
        case 'linear'
            parfor wid=1:N_WORKER
                Ktemp(:,:,wid)=X(:,:,wid)*X(:,:,wid)';
            end
            K=sum(Ktemp,3)*arg;
        case 'poly'
            parfor wid=1:N_WORKER
                Ktemp(:,:,wid)=X(:,:,wid)*X(:,:,wid)';
            end
            K=(sum(Ktemp,3)/arg(2)+1).^arg(1);
        case 'rbf'
            parfor wid=1:N_WORKER
                Ktemp(:,:,wid)=squareform(pdist(X(:,:,wid),
                 'euclidean'));
            end
            K=exp(-0.5*sum(Ktemp.^2,3)/arg^2);
        case 'sigmoid'
            parfor wid=1:N_WORKER
                Ktemp(:,:,wid)=X(:,:,wid)*X(:,:,wid)';
            end
            K= tanh(arg(1)*sum(Ktemp,3)+arg(2));
        otherwise
            disp('kernel not recognized')
            K=-1;
```

```
    end
else % multi kernel
    num_K=length(ker);
    Ktemp=zeros(n,n,num_K,N_WORKER);


    parfor wid=1:N_WORKER
        for nk=1:num_K
            switch ker{nk}
                case 'rbf'
                    Ktemp(:,:,nk,wid)=squareform(pdist(X(:,:,
                      wid),'euclidean'));
                case 'linear'
                    Ktemp(:,:,nk,wid)=X(:,:,wid)*X(:,:,wid)';
                case 'poly'
                    Ktemp(:,:,nk,wid)=X(:,:,wid)*X(:,:,wid)';
                case 'sigmoid'
                    Ktemp(:,:,nk,wid)=X(:,:,wid)*X(:,:,wid)';
                otherwise
                    error('wrong kernel type');
            end
        end
    end

    K=zeros(n,n,num_K);
    for nk=1:num_K
        switch ker{nk}
            case 'rbf'
                K(:,:,nk)=exp(-0.5*sum(Ktemp(:,:,nk,:).^2,4)/
                  arg{nk}^2);
            case 'linear'
                K(:,:,nk)=sum(Ktemp(:,:,nk,:),4)*arg{nk};
            case 'ploy'
                K(:,:,nk)=(sum(Ktemp(:,:,nk,:),4)/arg{nk}
                  (2)+1).^arg{nk}(1);
            case 'sigmoid'
                K(:,:,nk)=tanh(arg{nk}(1)*sum(Ktemp(:,:,nk,:),
                  4)+arg{nk}(2));
        end
    end

end
```

INDEX

Data-Variant Kernel Analysis, First Edition. Yuichi Motai.

Wiley Series on Adaptive and Cognitive Dynamic Systems

Editor: Simon Haykin

Adali and Haykin · *Adaptive Signal Processing: Next Generation Solutions*
Beckerman · *Adaptive Cooperative Systems*
Candy · *Model-Based Signal Processing*
Chen, Haykin, Eggermont, and Becker · *Correlative Learning: A Basis for Brain and Adaptive Systems*
Chen and Gu · *Control-Oriented System Identification: An H_∞ Approach*
Cherkassky and Mulier · *Learning from Data: Concepts, Theory, and Methods*
Cirrincione and Cirrincione · *Neural Based Orthogonal Data Fitting: The EXIN Neural Networks*
Costa and Haykin · *Multiple-Input Multiple-Output Channel Models: Theory and Practice*
Diamantaras and Kung · *Principal Component Neural Networks: Theory and Applications*
Haykin · *Unsupervised Adaptive Filtering: Blind Source Separation*
Haykin · *Unsupervised Adaptive Filtering: Blind Deconvolution*
Haykin and Liu · *Handbook on Array Processing and Sensor Networks*
Haykin and Puthussarypady · *Chaotic Dynamics of Sea Clutter*
Haykin and Widrow · *Least Mean-Square Adaptive Filters*
Hossain, Le, and Niyato · *Radio Resource Management in Multi-Tier Cellular Wireless Networks*
Hrycej · *Neurocontrol: Towards an Industrial Control Methodology*
Hyvärinen, Karhunen, and Oja · *Independent Component Analysis*
Kristić, Kanellakopoulos, and Kokotović · *Nonlinear and Adaptive Control Design*
Mann · *Intelligent Image Processing*
Motai · *Data-Variant Kernel Analysis*
Nikias and Shao · *Signal Processing with Alpha-Stable Distributions and Applications*
Passino and Burgess · *Stability Analysis of Discrete Event Systems*
Sánchez-Peña and Sznaier · *Robust Systems Theory and Applications*
Sandberg, Lo, Fancourt, Principe, Katagairi, and Haykin · *Nonlinear Dynamical Systems: Feedforward Neural Network Perspectives*
Sellathurai and Haykin · *Space-Time Layered Information Processing for Wireless Communications*
Spooner, Maggiore, Ordóñez, and Passino · *Stable Adaptive Control and Estimation for Nonlinear Systems: Neural and Fuzzy Approximator Techniques*
Tao · *Adaptive Control Design and Analysis*
Tao and Kokotovi⊠ · *Adaptive Control of Systems with Actuator and Sensor Nonlinearities*
Tsoukalas and Uhrig · *Fuzzy and Neural Approaches in Engineering*
Van Hulle · *Faithful Representations and Topographic Maps: From Distortion- to Information-Based Self-Organization*

Vapnik · *Statistical Learning Theory*

Werbos · *The Roots of Backpropagation: From Ordered Derivatives to Neural Networks and Political Forecasting*

Yee and Haykin · *Regularized Radial Bias Function Networks: Theory and Applications*

p 101 - kernel adaptation

~~102~~

103 - updating the Gram matrix

104 - GEVD

105 - online data (time index)

update time t → t+1

106 - kernel alignment determination

114 - alignment factors

120 - On kernel target alignment

123 - adaptability